A Tapestry of Values

A Tapestry of values

A TAPESTRY OF VALUES

An Introduction to Values in Science

Kevin C. Elliott

OXFORD
UNIVERSITY PRESS

Oxford University Press is a department of the University of Oxford. It furthers
the University's objective of excellence in research, scholarship, and education
by publishing worldwide. Oxford is a registered trade mark of Oxford University
Press in the UK and certain other countries.

Published in the United States of America by Oxford University Press
198 Madison Avenue, New York, NY 10016, United States of America.

Library of Congress Cataloging-in-Publication Data
Names: Elliott, Kevin Christopher, author.
Title: A tapestry of values : an introduction to values in science / Kevin C. Elliott.
Description: First edition. | New York, NY : Oxford University Press, [2016] |
Includes bibliographical references.
Identifiers: LCCN 2016026045| ISBN 9780190260811 (pbk. : alk. paper) |
ISBN 9780190260804 (hardcover : alk. paper) | ISBN 9780190260835 (ebook) |
ISBN 9780190260842 (online course) | ISBN 9780190260828 (pdf)
Subjects: LCSH: Research—Moral and ethical aspects. | Values.
Classification: LCC Q180.55.M67 E45 2016 | DDC 174/.95—dc23
LC record available at https://lccn.loc.gov/2016026045

9 8 7 6 5 4 3 2

Paperback printed by Webcom, Inc., Canada
Hardback printed by Bridgeport National Bindery, Inc., United States of America

For my children, Jayden and Leah

CONTENTS

CONTENTS

PREFACE

My motivation for writing this book stemmed from a gap I perceived in the literature on science and values. A host of recent books and articles have addressed this topic in one form or another. The fields of history and philosophy of science (HPS) and science and technology studies (STS) have focused a great deal of attention on the intersection of science and values. Moreover, semi-popular books like *Merchants of Doubt* (Oreskes and Conway 2010), *Doubt Is Their Product* (Michaels 2008), and *Bad Pharma* (Goldacre 2012) have made the wider public more aware of the ways that values influence science. Nevertheless, much of this literature suffers from two weaknesses. First, it sometimes emphasizes the general point that science is deeply influenced by values but without getting into nitty-gritty details about the precise ways in which values influence specific choices about scientific methods, concepts, assumptions, and questions. Second, it often highlights the presence of values in science without clearly exploring how to tell the differences between influences that are acceptable and those that are not.

The field of philosophy is well equipped to fill both these gaps. Philosophers of science have in fact written a good deal about both these issues—identifying the range of ways that values influence science and deciding which influences are appropriate and which are not. Unfortunately, most of this work has been written in the form of scholarly journal articles and technical books that are not accessible to a wide audience. My goal was to incorporate these philosophical insights into a book that a college freshman or an interested member of the general public could read and find to be helpful and informative. Also, while taking a philosophical perspective, I wanted the book to be sufficiently interdisciplinary to be appropriate for introductory courses on science policy, research ethics, history of science, environmental studies, and STS, as well as the philosophy of science. Finally, I wanted the book to be relevant to practicing scientists and policymakers.

With these goals in mind, I have written this book so that it focuses heavily on case studies of actual situations in which values have influenced scientific practice. In order to enhance the book's readability, I have also avoided using in-text citations and footnotes in favor of providing a list of sources after each chapter that can direct readers to the relevant references at the end of the book. While striving for readability, I have also maintained a philosophical stance that consistently asks whether and why particular value influences are justifiable. The book's chapters are organized around five different features of science that can be influenced by values: the choice of research topics (chapter 2), the manner in which a topic is studied (chapter 3), the aims of specific scientific investigations (chapter 4), the ways scientists respond to uncertainty (chapter 5), and the language employed for describing results (chapter 6). This enables me to provide an accessible introduction to the views of numerous prominent philosophers who have discussed the roles of values in scientific research, including Philip Kitcher and Janet Kourany (chapter 2), Elizabeth Anderson, Hugh Lacey, and Helen Longino (chapter 3), Kristen Intemann and Nany Tuana (chapter 4), Heather Douglas (chapter 5), John Dupré and Dan McKaughan (chapter 6), Kristin Shrader-Frechette (chapter 7), and many others. In keeping with the interdisciplinary character of the book, I am also able to touch on the views of major science policy and STS scholars, including Phil Brown (chapter 3), Roger Pielke Jr. and Daniel Sarewitz (chapter 5), and Steven Epstein, David Guston, Abby Kinchy, and Shobita Parthasarathy (chapter 7).

One disadvantage of the approach I have taken is that it does not allow me to provide an extensive philosophical justification of my own views about how to distinguish legitimate and illegitimate roles for values in science. In chapters 1 and 8, I provide a brief introduction to my own approach. I identify two scenarios in which values have a legitimate role to play in science. First, in many cases scientists are forced to make choices that are not completely settled by the available evidence but that serve some ethical or social values over others. Even if scientists are not deliberately trying to promote particular values when they make these choices, I contend that they should recognize the value-laden aspects of these decisions and take them into account. Second, I argue that values are justifiable in many contexts because they help scientists to achieve legitimate goals (e.g., doing their research in a manner that serves social needs and priorities).

When either of these two scenarios is present (i.e., when value influences are unavoidable or when they help to achieve legitimate goals), values have an appropriate role to play in science. In order to determine whether these scenarios are present and which values should be incorporated in scientific practice, however, additional conditions typically need to be met. I emphasize three conditions in this book: (1) scientists should be as *transparent* as possible about their data, methods, models, and assumptions so that value influences can be scrutinized; (2) scientists and policymakers should strive to incorporate values that are *representative* of major social and ethical priorities; and (3) appropriate forms of *engagement* should be fostered so that relevant stakeholders can help to identify and reflect on value influences. Some of these ideas are discussed in my previous work (e.g., Elliott 2011b, 2013a, b; Elliott and McKaughan 2014; Elliott and Resnik 2014; McKaughan and Elliott 2013). The motivation behind these conditions is that we should strive to support the most socially and ethically appropriate values possible, but there is bound to be disagreement about which values are best. Thus, it is important to make everyone aware of the values that are influencing science so that others can determine how their approach to science might differ based on their own values.

Fortunately, while my preferred approach differs from other philosophical work in some of its details, it agrees in its broad outlines with many other accounts about the major ways in which values can appropriately influence science. Therefore, my hope is that this book provides an accessible, effective introduction to the major philosophical views about how and why values can appropriately influence scientific research. Ultimately, I hope it also encourages thoughtful reflection about how we can guide the values that influence research so that we can better serve our ethical and social goals.

With this in mind, I would like to offer a few comments about the book's cover. It is an image from "The Unicorn in Captivity," the final tapestry in a seven-piece sequence known as the "Unicorn Tapestries." Housed at the Cloisters, a branch of the Metropolitan Museum of Art in New York City, these famous tapestries from the late Middle Ages describe the hunt for a unicorn. But the appropriateness of the image goes beyond the fact that it is a famous tapestry. The unicorn is a powerful symbol, and the sequence of tapestries is commonly interpreted to represent the courting of a bridegroom or the incarnation of Christ. Thus, it illustrates the values of love and religious commitment, which this book suggests are not as distinct from science as has often been assumed. I am also intrigued by the parallels between the hunt to capture and tame the elusive unicorn, and this book's account of the hunt to "tame science" so that it promotes our ethical and social values. Looking carefully at the tapestry, one can see juice from the fruits of the pomegranate tree falling on the unicorn's side. While typically interpreted as symbolic of the fruitfulness of marriage, perhaps in the context of this book they could also be interpreted as the fruits that we all hope science can generate for society.

Kevin C. Elliott
East Lansing, MI
May 2016

ACKNOWLEDGMENTS

I am very grateful to Michigan State University and to the Dean of Lyman Briggs College, Elizabeth Simmons, for the support I received to pursue this project. The core of my writing took place during the spring semester of 2015, when I had a course release from Briggs. Also, much of the motivation for pursuing this project stemmed from the focus on undergraduate education in Briggs and my desire to develop a textbook that would be useful for its introductory course in the History, Philosophy, and Sociology of Science. This book is much stronger because of the advice I received from the students who tested out the manuscript in my Fall 2015 Introduction to HPS course, my Spring 2016 STEPPS Capstone, and my Spring 2016 Graduate Seminar in Fisheries and Wildlife.

I am also grateful for the many colleagues with whom I have worked and studied. While I cannot remember exactly when I developed the idea for this book's title, I think it came from a conversation with Heather Douglas, who has also talked about scientific practice as a tapestry. My discussions with Heather have been invaluable for the development of my ideas about science and values. My friend Dan McKaughan has also had a major influence on this book, and readers will notice that many of the ideas in chapters 4 and 6 originated in articles that we wrote together. Chapter 4 also contains material I learned from writing an article with David Willmes. In addition, this book is full of insights I gleaned from Justin Biddle.

Janelle Elliott and Dan Steel read the entire manuscript and commented on it. Two anonymous reviewers for Oxford offered excellent suggestions on the original book proposal, and the reviewer for the penultimate version of the manuscript provided comments that improved the book significantly. I also gained important feedback about the book from Robyn Bluhm, Matt Brown, Erin Nash, Pat Soranno, the attendees of my presentation for the spring 2016 graduate conference on science and values at the University of Washington, and the participants at the 2016 meeting of the Consortium for Socially Relevant Philosophy of/in Science and Engineering. More broadly, I am grateful to the inspiration for thinking about science and values that Kristin Shrader-Frechette, Don Howard, and Janet Kourany provided at the

University of Notre Dame, and I am thankful for the countless conversations with other scholars that formed my thinking about these issues.

I am particularly grateful to my wife, Janet. I argue in this book that scientific practice is like a tapestry, composed of many different threads woven together. But my writing of this book was like a tapestry as well, insofar as my reading and writing and thinking were woven together with countless family activities. My work in this book reflects the many blessings I received from Janet and from our children, Jayden and Leah. I am dedicating this book to our children in hopes that it can help foster research efforts that make the world a better place for them and for their own children.

A Tapestry of Values

A Tapestry of Values

CHAPTER 1

An Introduction to Values in Science

On January 26, 1943, renowned Russian scientist Nikolai Ivanovich Vavilov died in a Soviet prison. His death is particularly horrifying, given that he starved to death after devoting his life's work to providing food for his country and the world. As a geneticist and agricultural scientist, he developed the insight that plant breeders should try to identify the geographical locations where key food crops evolved. In those regions, he suggested that scientists could find immense genetic diversity among the different crop varieties that still grew there. This genetic diversity could be used for breeding new crop varieties with valuable qualities, such as drought resistance, tolerance to extreme temperatures, and high yields. The search for improved varieties was crucial to the welfare of the Russian people because they were suffering perennial food shortages and famines. So, armed with his genetic insights, Vavilov set off across the world in search of seeds and samples.

Vavilov's story illustrates the problems that values can cause in science. Despite his work on behalf of the Russian people, he was ultimately sent to prison because Josef Stalin, the leader of the Soviet Union, became convinced that the genetic theory that undergirded Vavilov's work ran counter to the values of the Soviet leadership. Because of this clash of values, and because Stalin needed a scapegoat to blame for the failures of his collectivist agricultural program, he decided to suppress the field of genetics.

But values can also play very positive roles in science. After telling the story of Vavilov, this opening chapter recounts Theo Colborn's pioneering research on environmental pollution during the 1980s and 1990s. Colborn's story illustrates the many beneficial ways in which values can influence research. Together, the stories of Vavilov and Colborn highlight the importance of the two main questions that we will be exploring throughout this book. First, what are the major ways in which scientific reasoning can be influenced by values? Second, how can we tell whether those influences are acceptable or not? In preparation for answering those questions, the final sections of this chapter provide an overview of the book and clarify some of the philosophical ideas that we will explore throughout the following chapters.

VALUES IN THE STORY OF VAVILOV

In his efforts to develop agricultural breakthroughs that could help the Russian people, Vavilov ultimately launched more than two hundred expeditions around the globe and collected seeds from hundreds of thousands of plants. As recounted by his biographer, Peter Pringle, Vavilov's adventures were extraordinary. In 1916, he led an expedition to the Pamirs, a mountainous region on the border of Russia, Afghanistan, and China. Because the safest routes were too dangerous to pursue because of military conflicts, they had to go by horse over a treacherous glacier, and Vavilov and his horse almost fell to their deaths. In 1924, he guided the first Russian scientific expedition to Afghanistan, where he suffered from malaria and traveled through such dangerous areas that he struggled to find local guides willing to accompany him. In 1927, he traveled to Abyssinia (modern-day Ethiopia) and Eritrea. On that expedition, his team escaped from a group of armed men by gifting them with brandy and sneaking away while they were sleeping it off. On a trip through Spain, he was supposed to be followed by two police agents, but his schedule was so exhausting that they gave up following him during the day and agreed to catch up with him each night. He collected samples from around the world, including the United States, Canada, Western Europe, China, Japan, and Central and South America.

Vavilov's work was acclaimed both in the Soviet Union and internationally. In 1926, he received the Lenin Prize, the top Soviet award for scientific activity. In 1929, he became the youngest scientist ever to be elected as a full member of the USSR Academy of Sciences, and he became the director of the Lenin Academy of Agriculture. He was respected by the world's top geneticists, including Thomas Hunt Morgan and Hermann Muller. In Leningrad, he created the world's largest seed bank, which contained more than 250,000 samples, and he tested new plant varieties at 300 experiment stations that he oversaw across the Soviet Union. When the Nazis laid siege to Leningrad during World War II and the people of the city were starving, workers at Vavilov's institute faithfully guarded the seeds, and some of the workers even starved themselves rather than harming the collection. The passion with which these workers protected the collection illustrates the profoundly beneficial role that values can play in science.

Unfortunately, Vavilov's story is also frequently held up as a classic example of how values (in this case, political ideology) can harm science. It seems nearly inconceivable that a humanitarian and scientific hero of Vavilov's stature could end up starving to death in prison, but he was engaged in his plant breeding at the same time that Josef Stalin was collectivizing Soviet agriculture and repressing political dissidents. Stalin's policies aggravated the food shortages that already plagued the Soviet Union. In response, Stalin's administration placed increasing pressure on Vavilov to increase agricultural yields as quickly as possible, but Vavilov insisted that there were limits to the speed at which his breeding techniques could produce new advances. At the same time, an ambitious young scientist named Trofim Lysenko claimed to be able to make much swifter advances by pursuing different agricultural strategies.

According to Pringle, Lysenko was an example of what the Soviets called "barefoot scientists." These were researchers who had not graduated from a university and who focused on practical issues of direct relevance to the people. Beginning in 1927, the official newspaper of the communist party, *Pravda*, began drawing attention to Lysenko's work as a plant breeder. He claimed to be able to advance the yield of crops dramatically by altering the exposure of seeds to environmental factors like temperature, moisture, and light. Vavilov admired some aspects of Lysenko's work, but he and his fellow agricultural scientists also recognized that it suffered from errors, exaggerations, and perhaps fraud. Nevertheless, Lysenko's approach fit very well with the ruling party's Marxist philosophies that also emphasized the importance of environmental factors rather than inherited traits. Moreover, Lysenko was extremely strategic about using his peasant background and his lack of education to his advantage in a cultural setting that was extremely suspicious of the traditional intelligentsia.

Because Stalin was looking for scapegoats to blame for the failure of his collectivization program in the 1930s, Vavilov and his fellow geneticists became easy targets. For a long time, Vavilov attempted to promote peace with Lysenko, suggesting that his environmental approaches to agricultural science were complementary to the genetic breeding strategies that Vavilov pursued. Nevertheless, Lysenko chafed under the criticisms that his work received from many of Vavilov's fellow geneticists. As Lysenko gained more and more power in the Soviet hierarchy, he became increasingly aggressive about labeling the field of genetics as a false, Western, "bourgeois" approach to science. By 1938, Lysenko became the president of the Lenin Academy of Agriculture that Vavilov had led until his falling-out with the Soviet political establishment. Meanwhile, Stalin was engaging in vicious purges of vast numbers of people accused of being "counter-revolutionaries," and geneticists were increasingly placed into that category by the late 1930s.

While some geneticists were being executed, Vavilov remained free for the time being. He may have benefited from his international stature and Stalin's desire to avoid attracting unwanted attention to his reign of terror. Nevertheless, the Soviet leadership ultimately plotted to arrest Vavilov quietly while he was on an expedition to western Ukraine. On August 6, 1940, he was seized and taken to jail. He was interrogated nearly four hundred times over the next 11 months. Between August 10 and August 24, he was interrogated for 120 hours while being subjected to severe sleep deprivation. By the end of August, he confessed to the false charge of being a member of a "rightist" organization within the agricultural system, but he refused to confess to espionage despite unrelenting interrogations throughout the following year. Based on fraudulent testimony, he was ultimately convicted of espionage and sentenced to death. On appeal, he received a reduction of his sentence and was allowed to remain in jail, but the grim conditions of his incarceration led to his death a year and a half later.

The Soviet adoption and defense of Lysenkoism, and the accompanying persecution of scientists who held differing views, is frequently held up as a prime example of the dangerous consequences that can arise when values are allowed to influence science. If science is not kept pure of political, religious, and ethical values, so the worry goes, it runs the risk of being hijacked by ideologies that prevent scientists from arriving at the truth. This criticism of Lysenkoism may not be entirely fair, however. Lysenko was often labeled a "pseudoscientist" during the middle decades of the twentieth century, but more recent scholarship has shown that the situation was somewhat more complicated. While his efforts to abandon traditional genetics were poorly defended, his initial research showing that exposure to low temperatures could hasten the development of crops was well received by the scientific community. Moreover, his research was informed by the goal of integrating theoretical work with practical concerns, which was an important priority of the Soviet Union at that time. The following chapters argue that under the right conditions, it can be acceptable for science to be informed by these sorts of practical aims.

Thus, the story of Vavilov and Lysenko appears to be problematic not solely because values played a role in Lysenko's research but rather because the case failed to meet additional conditions that are important for incorporating values in a legitimate fashion. The most obvious problem was that many scientists who resisted Lysenko's favored approaches were brutally repressed. Those who challenged his views about genetics were often imprisoned or killed, and even those who were not arrested were afraid to provide critical feedback about Lysenko's work. As a result, scholars were not able to engage critically with his research and discuss its quality. Another worry about Lysenko's work is that, perhaps because of his limited training, his experiments and results were not described carefully. This lack of transparency made it even more difficult for the scientific community to evaluate it. We will return throughout this book to the importance of both critical engagement and transparency if values are to be incorporated appropriately in scientific research.

Consider another, more recent story involving values in science. When Theodora (Theo) Colborn passed away in December 2014, she was hailed for launching a revolution in our understanding of environmental pollution. She was frequently compared to the pioneering environmentalist Rachel Carson because of their similar efforts to highlight the hazards associated with toxic chemicals. It is therefore fitting that Colborn received numerous awards named after Carson, as well as a variety of other honors, including the Blue Planet Prize, which is often regarded as the Nobel Prize for the environmental sciences.

Colborn's eventual impact would have seemed highly unlikely in 1978, when she was a 51-year-old retired pharmacist who decided to embark on a new career path. She was living near Rocky Mountain Biological Laboratory in Colorado and had been motivated by her love of birds to help volunteer for environmental organizations. In order to develop more compelling credentials, she decided to pursue a master's degree in ecology, and she went on to earn a PhD in zoology from the University of Wisconsin–Madison. After she finished her degree in 1985, she received a fellowship to work for the US Office of Technology Assessment in Washington, DC. In 1987, she received a position at a nonprofit think tank called the Conservation Foundation to study the effects of environmental pollution in the Great Lakes region.

At that time, scientists were especially focused on the concern that toxic chemicals released into the environment were causing cancers in humans and wildlife. Nevertheless, she could not find compelling evidence that people living in the Great Lakes region were suffering from abnormally high rates of cancer. She did find, however, that the animals living in the region were experiencing a wide range of surprising abnormalities. For example, in some herring gull colonies, scientists were finding two females in a nest rather than a male and a female, apparently because of a shortage of males. Researchers also observed strange parental behavior in various bird species—they seemed less interested than normal in defending their nests and incubating the eggs. Many birds were also born with deformities, and other animals, such as mink, were having trouble reproducing. Colborn recognized that something was wrong, but these findings did not fit the cancer paradigm within which most scientists were working.

Colborn ultimately synthesized a number of findings and led the development of a new paradigm for approaching environmental pollution. She realized that many of the problems faced by organisms in the Great Lakes were related to reproduction and development, especially in the offspring of adults exposed to pollutants. Building on experimental work performed by other scientists, she proposed that many chemicals were generating harmful effects by interfering with animals' hormonal systems. Because the hormonal system is deeply intertwined with the immune system and the neurological system as well as the process of development, environmental pollutants could cause a wide array of problems by altering hormones. In some cases, the harmful effects could involve cancers, but a wide variety of other consequences could also result.

These harmful health effects (which Colborn called "endocrine disruption") have raised a number of new concerns for scientists and policymakers. Because organisms are sensitive to extremely low levels of hormones, especially during sensitive periods of fetal development, some scientists worry that environmental pollutants acting in this manner could cause problems at much lower dose levels than previously thought. The effects of endocrine disruptors are also difficult to study; they may generate a number of subtle consequences that are more difficult to recognize than cancer, they may cause different problems at low doses than at higher doses, and they may generate effects only when organisms are exposed to them at crucial "windows" of development. Some scientists worry that humans are already experiencing harmful effects from exposure to endocrine-disrupting chemicals. These may include birth defects, infertility, weakened immune systems, attention-deficit disorders, decreased male sperm counts, and cancers.

What is particularly noteworthy for the purposes of this book is that values influenced Colborn's pioneering research on endocrine disruption in a variety of ways. First, her discovery of the phenomenon was due in large part to her passion for protecting the environment. She would not have pursued a new career as an environmental scientist—let alone engage in hours of detective work to pore over research articles on the plight of Great Lakes wildlife—if it were not for her strong environmental values. Then, she worked with others to write a popular book, *Our Stolen Future*, which drew attention to the potential hazards associated with endocrine-disrupting chemicals. Because of the authors' strong concerns about public health, they thought it was important to give people clear warnings about the potential threats they faced. Colborn's concerns about public welfare ultimately drove her to found an international nonprofit organization, The Endocrine Disruption Exchange (TEDX), in order to facilitate the compilation and dissemination of information about endocrine disruptors.

In contrast to the case of Lysenko, values appear to have played a largely positive role in Colborn's research. Nevertheless, her work has not entirely escaped controversy. Critics have complained that she sometimes leaped ahead of the scientific evidence and drew stronger conclusions than they thought the evidence warranted. In particular, her book *Our Stolen Future* was criticized for being too aggressive about drawing the conclusion that humans were being harmed by the levels of endocrine-disrupting chemicals currently present in the environment. The critics worried that it was irresponsible to arouse public concerns while the evidence was still highly uncertain.

Nevertheless, Colborn and her coauthors insisted that it was important to make people aware of the dangers they could be facing.

We will return in chapter 5 to the contention that Colborn's environmental values may have caused her to be too bold about drawing conclusions on the basis of limited evidence. For now, we can make two observations. First, both Colborn's and Vavilov's stories illustrate the importance of thinking more carefully about when value influences are appropriate and when they are not. Second, even though the influences of values in Colborn's case are complex, they appear to have played many valuable roles. Her passionate concerns for public and environmental health helped her to launch a revolution in our understanding of chemical safety. And even in the more controversial elements of her work, such as the writing of her popular book, the influences of values appear to be much more justifiable than in the case of Lysenko. For example, she and her co-authors tried to acknowledge the weaknesses and limitations of their interpretations in *Our Stolen Future*, and they appealed to high-quality published research that could be accessed and evaluated by others. Moreover, her critics were not under social pressure to hide their views; instead, prominent journals published warnings about the limitations of the book. Thus, the conditions of transparency and critical engagement were met to a much greater extent in the case of Colborn than in the case of Lysenko.

OVERVIEW OF THE BOOK

This quick sketch of two episodes in scientific research illustrates the importance of thinking more deeply about the roles that values play in science. Values can influence science in a variety of ways, including by providing the inspiration to pursue particular topics, by altering the questions and methods that researchers pursue, and by changing the amount of evidence they demand before drawing conclusions. But it is also clear that value influences can be problematic. They can lead to the suppression of important ideas, they can result in questionable interpretations of the available evidence, and they can cause researchers to mislead the public about the state of scientific research. In response to these challenges, this book focuses on two questions. First, what are the major ways in which scientific reasoning can be influenced by values? Second, how can we tell whether those influences are appropriate or not?

These questions have preoccupied historians, philosophers, and sociologists of science in recent years. Some scholars have promoted the "value-free ideal," which is the notion that values should be excluded from central aspects of scientific reasoning, such as decisions about which methodologies or standards of evidence to employ. For example, the famous twentieth-century sociologist Robert Merton proposed that one of the central norms of science is "disinterestedness," according to which scientists should strive to avoid being influenced by personal, emotional, or financial considerations. In recent years, concerns about allowing values to influence science contributed to the "Science Wars." These were debates at the end of the twentieth century in which a variety of scientists worried that scholars in the social sciences and humanities were abandoning the notion that scientists could or should strive to arrive at objective, value-free truths.

The potential for values to damage science remains an important issue today, as members of both political parties worry about the politicization of science. Much climate-change denial appears to be fueled by conservative political values. In other words, those who are politically opposed to "Big Government" tend to question the existence of environmental problems like climate change because these problems appear to require government-led solutions. Similarly, conservative religious values appear to fuel a great deal of skepticism about evolutionary theory. And questionable influences of values on science are not limited to those on the political right. Skepticism about scientific evidence regarding the safety of vaccines and genetically modified foods appears to be influenced in at least some cases by other personal, social, or ideological values.

While some contend that the best solution to these problems is to hold fast to the value-free ideal, others contend that it is unattainable and detrimental both to science and society. They insist that values often have an important role to play in deciding what to study, how to study it, what to aim for, when to draw conclusions, and how to describe the findings. This book comes down on the side of those who reject the value-free ideal. It argues that social and ethical values are essential and unavoidable in many central aspects of scientific reasoning. When these legitimate influences of values are ignored, science still ends up being value-laden, but the influences of values are not subjected to adequate scrutiny or discussion.

Excluding values from scientific reasoning leaves the scientific enterprise severely weakened and prone to the influences of hidden values that are not thought through carefully. Attempting to exclude values is a bit like claiming that knives should no longer be allowed in kitchens because people could be injured by them. Values can cause serious problems in science, just like knives can cause significant injuries, but the fact that they can be used unwisely or inappropriately in some cases does not mean that they are problematic under all circumstances. Values have important roles to play in scientific reasoning; the key is to recognize the variety of ways in which they can exert their influences and to figure out when those influences are appropriate and when they are not.

In order to investigate these questions, each of the next five chapters explores a particular way in which values can influence science. Chapter 2 explores how values can legitimately influence the *choice of topics* that scientists investigate. We have already seen that Colborn's concerns about public health contributed to her enthusiasm for studying environmental pollution, and both Vavilov and Lysenko were motivated by the goal of alleviating hunger. Nevertheless, as we will see, it is often difficult to achieve research that accords with our ethical and social priorities. Chapter 2 explores some of these challenges and the efforts that have been made to address them.

Vavilov's and Lysenko's conflicting approaches to agricultural research illustrate that even when scientists are motivated to address the same general topic, they can engage in very different forms of research. Chapter 3 argues that *choices about how to study a topic* can be influenced (either explicitly or implicitly) by values. For example, some contemporary agricultural researchers focus on developing genetically engineered seeds, while others are more interested in developing ecologically friendly strategies for raising multiple crop and animal species together. Each approach has its own strengths and weaknesses that tend to promote different social values. Chapter 3 explores how the methods that scientists use, the assumptions that they employ, and the specific questions that they ask can all be value-laden.

Chapter 4 argues that values can also play an important role in *determining the aims of scientific inquiry in particular contexts.* This is important because scientists often have to weigh a variety of theoretical and practical goals when developing new theories, methods, or models. For example, when scientists are working with regulators or policymakers, they are sometimes asked to develop methods or models that generate results relatively quickly and inexpensively. In these situations, they sometimes have to make value-laden decisions about when it is appropriate to sacrifice detail and accuracy in order to achieve these other goals. In other cases, optimizing a model to predict some aspects of a phenomenon makes it more difficult for the model to predict other aspects of the phenomenon; thus, scientists may have to reflect on their overarching aims in order to decide what is most important for them to model accurately.

Chapter 5 examines the roles that values play in *decisions about how to respond to scientific uncertainty.* We have already encountered the worry that Colborn was too quick to draw the conclusion that endocrine-disrupting chemicals were harming human health. This is just one of many prominent examples in which scientists have been forced to decide when the available evidence was sufficient to warn the public of important threats. Values play an important role in deciding how bold or cautious to be in these situations. Chapter 5 also explores how values can sometimes contribute to uncertainty, insofar as groups of people with strongly held values are often motivated to generate research that supports their preferred positions on debated issues. In some cases this is acceptable, but in other cases it is problematic, and we will reflect on the differences between these cases.

Chapter 6 explores how values can play a role in *decisions about how to communicate and frame scientific information.* For example, philosopher Sheldon Krimsky has argued that one of the most significant aspects of Colborn's work was to show that a wide variety of abnormal effects in humans and wildlife could all be conceptualized as elements of a single new biological phenomenon: endocrine disruption. By developing this unifying concept, Colborn made it much more likely for people to take action to address it. But scientists have still debated what to call it. When a US National Academy of Sciences panel issued a report about endocrine disruptors, it decided to refer to these chemicals as "hormonally active agents" instead. Some authors of the report insisted that the word "disruption" sounded overly emotional. Moreover, they worried that it could scare people into thinking that chemicals inevitably cause harm whenever they affect an organism's hormonal system. Chapter 6 examines how these sorts of values can influence the description of scientific findings.

In the process of scrutinizing all these ways in which values influence science (see table 1.1), we will explore why some influences appear to be more justifiable than others. Over the course of chapters 2 through 6, we will find that three conditions appear to be particularly important for bringing values into science in an appropriate fashion: (1) value influences should be made as *transparent* as possible; (2) they should be *representative* of our major social and ethical priorities; and (3) they should be scrutinized through appropriate processes of *engagement* between different scholars and other stakeholders. Chapter 7 focuses especially on the third condition, asking how we can *promote engagement* between the scientific community and other segments of society about the values that influence science. This condition is particularly

Table 1.1 AN OVERVIEW OF THE WAYS THAT VALUES ARE RELEVANT
TO SCIENTIFIC RESEARCH

Roles for Values	Examples
Choosing topics to study	Putting significant social resources into studying endocrine-disrupting chemicals
Deciding how to study a particular topic	Deciding between different methods for agricultural research
Determining the aims of scientific inquiry in particular contexts	Determining which facets of a phenomenon are most important to model
Determining how best to respond to uncertainty	Choosing whether or not to alert the public that endocrine-disrupting chemicals could be causing harm to humans
Deciding how to describe, frame, and communicate results	Choosing whether to use the term "endocrine disruptor" or "hormonally active agent"

fundamental because it often plays a central role in achieving the other conditions of transparency and representativeness. It is important to remember, however, that these three conditions can be met to varying extents, and different people can disagree about how well they have been met. Thus, there can be disagreements about the extent to which particular value influences on science are truly appropriate or not.

Finally, chapter 8 provides an overview of the book's major lessons about how values can influence science and when those influences are justifiable. It also responds to several potential objections. One particularly important objection is that the book might seem to overstate the role that values play in science as a whole. For example, one might grant that values play a significant role in studying policy-relevant topics like agriculture and environmental pollution, but one might think that they are unlikely to influence theoretical research in areas like particle physics or cosmology. The contention of this book is that values are not completely absent from any area of science. Even in theoretical areas of physics, scientists and policymakers still face decisions about how much money to spend on different topics and how best to frame and communicate new findings. For example, when public funds have provided the support to pursue high-profile research projects, physicists face the same sorts of decisions that Colborn faced about when they have enough evidence to inform the public about new findings. The following chapters use examples from a number of different fields of science to show that the issues discussed in this book are not unique to a few subject areas. It is also worth noting that many of the fastest growing areas of research involve applied topics that have significant ramifications for society. These are precisely the areas of science that are most likely to be permeated with value judgments.

CLARIFYING AND JUSTIFYING VALUES

In an effort to be accessible to a variety of different readers, this book focuses heavily on concrete examples, and it tries to minimize complicated conceptual arguments. However, in order to prepare for the subsequent chapters, it will be helpful to offer a few preliminary clarifications about the nature of values before moving on. Those who find such philosophical details particularly tiresome may choose to move quickly on to the following chapters and return to the latter part of this chapter only if they desire further clarification.

Broadly speaking, a value is something that is desirable or worthy of pursuit. So, for example, scientists typically value accurate predictions, clear explanations, logical consistency, fruitful research projects, honesty, credit for their accomplishments, health, economic growth, environmental sustainability, and global security. Of course, not all scientists value each of these qualities to the same extent, so we will see that they often have to deliberate about which ones are most important to prioritize in particular circumstances. This array of values can be classified in several ways. Some values can be categorized based on their origin or source; for example, we often talk about "religious" values. Sometimes values are categorized based on their topic area (e.g., "political" values). In other cases, values are categorized based on what they help us to achieve. For example, some things that are valued, such as accurate predictions and logical consistency, are often called "epistemic values," because they are regarded as contributing to the goal of gaining knowledge. (The word "epistemic" comes from the Greek word "episteme," meaning knowledge.) Other values are often called "non-epistemic values," insofar as they do not consistently help us gain knowledge. For example, the fact that a theory fits well with one's preferred political values seems irrelevant to deciding whether the theory is true, whereas finding out that the theory generates accurate predictions does, all else being equal, increase the likelihood that it is true.

When most people use the term "value," they are usually thinking about ethical or political or religious values that are typically regarded as non-epistemic. Therefore, this book will use the term "values" in this popular sense. It will also refer to "value judgments" or judgments that are "value-laden." Value judgments are scientific choices that cannot be decided solely by appealing to evidence and logic. As the subsequent chapters illustrate, these judgments include decisions about methods, assumptions, interpretations, and terminology.

Sometimes these value judgments are consciously influenced by values, but in other cases they just support some values over others. For example, when Theo Colborn decided that there was adequate evidence to inform the public about the potential hazards of endocrine-disrupting chemicals, she was making a value judgment. Even if she was not purposely making this decision with the goal of supporting a particular set of values, her choice served the value of promoting public health over alternative values, such as promoting the short-term economic growth of the chemical industry. In many cases, individual scientists do not even have much control over the value judgments that inform their work; they may need to study particular topics or ask particular questions because that is what they are paid to do. Whether value influences are intentional or unintentional, and whether they are generated by individuals or institutions, our goal throughout this book is to explore when they have legitimate roles to play in science.

But how can we go about justifying roles for values in science? One possible strategy would be to show that these values actually do play a role in research. For example, one might show that scientists are typically influenced by a range of personal, social, and financial concerns. But this is a bit too simple. Scientists, policymakers, and political leaders often do things that we do not think are really acceptable or legitimate. For example, Stalin was apparently motivated by Marxist values (as well as political motivations to stay in power) when he suppressed the science of genetics, and almost no one would say that it was acceptable to engage in that sort of repression. Similarly, chapter 5 shows that some of the prominent scientists who have raised doubts about climate change appear to have been motivated by conservative political values to raise misleading objections about the work of other scientists. It does not seem appropriate for these political values to skew scientific discourse in this manner.

Some philosophers have coined a phrase, "the problem of wishful thinking," to describe these sorts of activities. At its core, wishful thinking consists in accepting or rejecting a hypothesis or theory just because one wants it to be true or false. It is probably relatively rare to commit wishful thinking in such a blatant way—most people typically think they have good reasons for their views. But there are a variety of other activities that are closely related to wishful thinking. These include using "rigged" methods that generate predetermined outcomes, ignoring evidence that conflicts with one's preferred conclusions, and repeating objections over and over even after they have been addressed. The notion that scientific reasoning should be value-free is often based on the worry that allowing values to influence science will result in wishful thinking. But we will see that this is not the case; we can acknowledge important roles for values in scientific reasoning while insisting that it is unacceptable to draw conclusions just because one wants them to be true.

Therefore, we need a different strategy for deciding when values have a legitimate role to play in scientific practice; we cannot merely point to examples where values have in fact influenced scientists. We need to decide which examples involve appropriate influences of values and which examples are problematic. Throughout this book, we will encounter two major justifications for bringing values into science. First, we will see that it is virtually impossible to prevent science from being influenced by values or supporting some values over others when making particular types of decisions about interpreting evidence or drawing conclusions. Therefore, given that all people—including scientists—have responsibilities to address the effects of their choices on other people, scientists should consider the roles of values in their work thoughtfully and intentionally rather than making value-laden decisions carelessly and thoughtlessly. For example, we will see in chapter 6 that it is often unrealistic to try to identify terms or ways of framing scientific information that are completely "value neutral"; all the available terms and frames have subtle connotations that support one value perspective or another. In cases like this, the problem is not just that it is difficult for scientists to avoid being influenced by their personal values; rather, regardless of the motivations of the scientists, their choices support some social values while weakening others. In these cases, responsible scientists should strive to recognize these value-laden choices and attempt to make these decisions as thoughtfully and transparently as possible.

This first justification might make it seem like values are just a "necessary evil" in science, but the second justification for values shows that they have a more positive role to play. The second major reason for bringing values into science is that they can help scientists to achieve legitimate goals. Political or social or religious values are typically regarded as problematic because they seem to be irrelevant for advancing the fundamental goal of science: obtaining well-justified, reliable information about the world. The fact that a particular hypothesis supports our political or religious preferences typically does not make it more likely to be true; this is why the problem of wishful thinking is a genuine problem. Nevertheless, the following chapters show that scientists have a number of other legitimate goals alongside the goal of obtaining reliable information about the world, and values are relevant to achieving many of these other goals.

What sorts of goals might these be? They include efforts to solve important social problems, to obtain information that addresses social priorities, to communicate information to the public in a responsible manner, and to inform policymakers in ways that benefit society. Some critics might point out that most of these goals incorporate a questionable assumption, namely, that scientists should be aiming to benefit society. In my view, however, this is a very reasonable assumption. Surely everyone, including scientists, should strive to help others if it is relatively easy to do so, and everyone has responsibilities not to carelessly or negligently harm others. And given that we as a society provide scientists with a great deal of financial and institutional support, it would be surprising if scientists did not have at least some responsibilities to do their work in a manner that benefits society. Thus, we will find that values have a legitimate role to play in many aspects of science because they help scientists to achieve their goals of serving society.

Whether value influences are inevitable or whether they help scientists to achieve their legitimate goals, additional conditions typically need to be met in order to make the influences of values fully justifiable. Indeed, one of the reasons for these additional conditions is that they help to determine when value influences truly are inevitable and whether the goals set by scientists in particular contexts truly are legitimate. As noted previously, this book focuses attention on three conditions: transparency, representativeness, and engagement. First, scientists should be as *transparent* as possible about their data, methods, models, and assumptions so that others can identify the ways in which their work supports or is influenced by particular values. Second, scientists and policymakers should strive to incorporate values that are *representative* of major social and ethical priorities. When clear, widely recognized ethical principles are available, they should be used to guide the values that influence science. When ethical principles are less settled, science should be influenced as much as possible by values that represent broad societal priorities. Third, scientists, citizens, and policymakers should encourage appropriate forms of *engagement* between scientists and other stakeholders. This engagement helps to facilitate both of the other conditions—transparency and representativeness—by promoting thoughtful scrutiny of values in science.

It is worth emphasizing that scientists can incorporate values into their work under these conditions without sacrificing scientific objectivity. The perspective taken by this book is that scientific objectivity is typically easier to maintain by acknowledging the roles of values in science than by trying to eliminate values. As the following chapters emphasize, scientists are forced to make all sorts of decisions as they go about their work. These include choices about what assumptions and methodologies to employ, what sorts of models to develop, what statistical techniques to use, and how to describe scientific results. These choices often serve some values over others, but the significance of these decisions often goes unnoticed. The goal of this book is to highlight the ways in which values are relevant to making these decisions so that scientists can be more transparent and generate more thoughtful deliberation about them. This willingness to make important judgments more explicit and to subject them to critical scrutiny is a hallmark of objectivity.

One final point worth mentioning is that this book will not be able to cover all the roles that values can legitimately play in scientific research. In fact, the focus here is primarily on roles for values that are somewhat surprising or subject to debate. Everyone already knows that ethical values are relevant to science in many obvious ways. For example, scientists are clearly expected to minimize the suffering of their experimental animals and to protect the rights of human research subjects. They are also expected to treat their colleagues and students with respect, to mentor their students conscientiously, to avoid plagiarizing the results of others, and to follow a number of other codes and regulations designed to protect people from harm. These influences of values are discussed extensively in textbooks on research ethics. We will focus here on roles for values that are more subtle and debatable. Rather than focusing on aspects of science that are somewhat peripheral, most of the avenues for values discussed in this book involve various facets of scientific reasoning. These roles for values are less widely appreciated, more controversial, and particularly fascinating to explore.

CONCLUSION

If we take seriously the lessons of this book—in particular, the conclusion that the value-free ideal needs to be abandoned—then we have important work to do. In particular, we need to explore ways to foster transparency, representativeness, and engagement in scientific practice. Fortunately, in the following chapters we will encounter many strategies for meeting these conditions. Some of the most intriguing strategies involve efforts to foster interdisciplinary collaborations between scientists and other scholars with diverse perspectives, as well as efforts to link scientists with community members who have a stake in their research. For example, chapter 3 discusses a particularly engaging case in which several dozen residents of Woburn, Massachusetts, sat down with Episcopal priest Bruce Young to discuss a surprising number of childhood leukemia cases in their neighborhood. They became convinced that the apparent cancer cluster that they had identified merited further scrutiny and that it might be related to water pollution. They urged the Massachusetts Department of Public Health (DPH) to investigate their concerns further, but a government report found the evidence to be ambiguous. Some of those citizens eventually convinced researchers at the Harvard School of Public Health to investigate their case further, and after conducting a study that included community volunteers they found that the cancers did appear to be related to pollution. The problem was eventually tracked down to city wells that were contaminated with chemicals from industrial facilities owned by W. R. Grace and Beatrice Foods, and the citizens ultimately obtained a financial settlement from W. R. Grace.

The Woburn case is now famous as an early example of community-based participatory research (CBPR). In CBPR, citizens work with researchers to design scientific studies in ways that address their concerns and meet their needs. CBPR is a natural outgrowth of the realization that research on chemical pollution and other policy-relevant issues incorporates a host of value judgments. The assumptions that scientists make, the specific questions that they ask, the methods that they employ, the standards of evidence that they demand, and the terms and concepts that they use for communicating their findings can all be influenced by implicit values. Citizens can help bring these values to light and suggest ways of steering research in directions that best fit their own concerns. One of the goals of this book is to advance these sorts of initiatives. If scientists, citizens, policymakers, and students are more sensitive to the value judgments that pervade science, they can address them in more thoughtful, socially beneficial ways.

SOURCES

For more information about Vavilov and his work as a plant researcher, see Graham (1993), Nabhan (2009), and Pringle (2003, 2008). Lysenko's work is discussed in Gordin (2012), Graham (1993), Pringle (2008), and Roll-Hansen (1985). The phenomenon of endocrine disruption and Colborn's work on it is described in Colborn, Dumanoski, and Myers (1996), Elliott (2011b), and Krimsky (2000).

For more on Merton's discussion of scientific norms, see Merton (1942). Koertge (1998) provides an analysis of the Science Wars. Discussions about the roles of values in studying chemical pollution can be found in Cranor (1990, 1993), Douglas (2009), and Elliott (2011b). For an introduction to the concepts of values, value judgments, and epistemic values, see Douglas (2009, 2015), Elliott (2011b), Elliott and Steel (2017), Longino (1990), and Steel (2010). Arguments for incorporating values in science based on their contributions to scientists' goals can be found in Anderson (2004), Brigandt (2015), Elliott and McKaughan (2014), and Hicks (2014). The problem of wishful thinking is discussed in Anderson (1995, 2004) and Brown (2013). Arguments for the value-free ideal can be found in Betz (2013), Hudson (2016), and Lacey (1999). The conditions of transparency, representativeness, and engagement are discussed further in chapter 8. Harding (2015) emphasizes that one can abandon the value-free ideal while continuing to promote scientific objectivity. The Woburn case is discussed in Brown and Mikkelsen (1990) and in Harr (1995). Citizen involvement in research is discussed more extensively in chapters 3 and 7 of this book.

CHAPTER 2

What Should We Study?

On February 21, 2006, Lawrence "Larry" Summers announced that he would resign from his position as president of Harvard University at the end of the school year. One of the major causes for his resignation was the furor caused by a speech in which he suggested that women might not have the same aptitude for math and science as men. As we will see, his speech constitutes just one episode in a long controversy over research that explores differences in cognitive abilities between men and women and between people of different races. While thousands of studies have been performed, it is doubtful that much of this research has been helpful for society, and there are reasons for thinking that it has actually been harmful. Thus, this case illustrates how social values—for example, our desire to promote equal opportunities for all citizens, both men and women—can become relevant to deciding which areas of research to prioritize.

At first glance, it might seem obvious and even trivial that values are important for deciding what research topics we as a society want to pursue. But while this may be somewhat obvious in principle, it turns out that we face a variety of fascinating questions about how exactly our values should play this role. In this chapter, we will explore three major ways in which values play a role in choosing research topics. First, our values may help us to make challenging decisions about which research topics to prioritize over others. Second, our values have important roles to play in guiding public funding for scientific research, which raises further questions about how best to do this. Third, values are important for evaluating the research funded by the private sector and looking for ways to influence this research so that it best serves our ethical and social goals.

PRIORITIZING RESEARCH: DIFFERENCES
IN COGNITIVE ABILITIES

The story of Larry Summers illustrates how our values can help us to prioritize some research topics over others. Born in 1954, he is the child of two economists. Amazingly, he is also the nephew of two Nobel Prize winners. His father's brother, Paul Samuelson, received the Nobel Prize in 1970 and authored one of the most important economics textbooks of the twentieth century. His mother's brother, Kenneth Arrow, won the Nobel Prize in 1972 and is famous for launching modern social choice theory (the study of how individual opinions or preferences can be combined into collective decisions). Summers initially planned to study physics but ultimately pursued economics himself. In 1983, at the age of 28, he became one of the youngest professors ever to be tenured at Harvard.

Summers's career is full of great accomplishments as well as significant conflicts. In 1991, he became Chief Economist at the World Bank. In 1993, he began working in the US Department of the Treasury in the Clinton administration, and in 1999 he became Secretary of the Treasury. While he was at the World Bank, a controversy arose because he signed a memo stating that more pollution should be exported from developed countries to developing ones because the lower wages in developing countries decrease the economic costs of pollution-related injuries and deaths. Summers has contended that the memo was written as sarcasm and that it was taken out of context.

Summers faced several major controversies during his time as president of Harvard. First, he clashed with the famous African American Studies scholar Cornel West. According to West, Summers accused him of missing too many classes, engaging in grade inflation, and embarrassing the university by producing a rap CD. As a result, West departed Harvard for Princeton. Summers also came under fire for his actions in support of his friend Andrei Schleifer, a fellow Harvard economist. Schleifer violated US conflict-of-interest guidelines by investing in Russian securities while helping to design Russia's privatization program as it shifted to a capitalist economy during the 1990s. When Schleifer and Harvard were sued by the federal government as a result, many faculty members accused Summers of inappropriately shielding Schleifer.

Summers made his infamous remarks about gender differences in cognitive abilities in January 2005 at a conference on developing a more diverse workforce in science and engineering. He suggested several hypotheses to explain why women were underrepresented in science and engineering positions at elite institutions. One hypothesis had to do with the difficulties of reconciling family demands with intense work schedules, a second hypothesis had to do with different aptitudes, and a third dealt with socialization and discrimination. While he suggested that the difficulties of reconciling work and family responsibilities (the first hypothesis) was probably the most important factor, he noted that it was also important to consider the fact that girls are less likely than boys to receive top scores on science and math tests during high school. He argued that new research in behavioral genetics indicated that these differences might be due in part to innate differences in ability rather than socialization.

Summers clarified during the speech that his comments were intended to be provocative, and he subsequently apologized and emphasized that he was fully committed to promoting opportunities for women in science and engineering. Nevertheless, a number of critics argued that his interest in innate differences in cognitive abilities was unwarranted and likely to discourage aspiring female scientists, especially given that there is so much evidence that socialization and discrimination are important factors undermining women's performance. Trying to evaluate all the evidence in this area would go beyond the scope of this book, but we can use Summers's speech as an opportunity to think about the ways that values can play a role in prioritizing research topics. In particular, ethical and social values can be used as a justification for making further studies of gender (or racial) differences in cognitive abilities a low social priority.

To make the case that these studies should be a low priority, it is helpful to consider an excellent overview by philosopher Janet Kourany of historical efforts to justify women's inferiority through scientific research. She notes that in the seventeenth century, women's brains were regarded as too cold and soft to compete with men's intellectual capabilities. In the eighteenth century, scientists argued that women's cranial cavities were too small to house the more effective brains that men had. In the nineteenth century, experts argued that if women engaged in too much intellectual work, it would harm their reproductive capabilities. In the twentieth century, it was argued that the lesser specialization of the two hemispheres in women's brains lessened their visuospatial skills.

Kourany notes that scholars continue to propose a variety of explanations for gender differences in cognitive abilities:

> That women's brains are smaller than men's brains, even correcting for differences of body mass; that women's brains have less focused cortical activity (lower "neural efficiency"); that women's brains have lower cortical processing speed (lower conduction velocity in their white matter's axons); and so on.[1]

The latest development is to map brain activity using fMRI (functional magnetic resonance imaging). Overall, it appears that more than 15,000 "human cognitive sex difference" studies were performed between 1968 and 2008. The

1. Kourany 2010, 5.

enthusiasm for this area of research does not appear to be slowing, given that more than 4,000 of these studies occurred between 1998 and 2008.

One of the valuable aspects of learning about this history is that it encourages caution when interpreting and evaluating studies in this area. For hundreds and even thousands of years, people have been attempting to justify their cultural assumptions that some gender or racial groups are inferior to others. Looking back at this history, we can see that there is serious danger that people will interpret complex bodies of scientific evidence in ways that fit their previously established assumptions rather than allowing the evidence to challenge their previously conceived notions. Moreover, even if there were compelling evidence for particular differences in cognitive abilities, the chances of these findings being misused to defend questionable public policies is huge. On this point, Kourany quotes a paper by Steven Rose from the journal *Nature*. He warns that people often justify cognitive differences research based on the argument that it could be used to assist "less well-endowed" groups to achieve greater success. But Rose argues that "in practice, claims that there are differences in intelligence between blacks and whites, or men and women, have always been used to justify a social hierarchy in which white males continue to occupy the premier positions (whether in the economy in general or natural science in particular)."[2]

Making Cognitive Differences a Low Research Priority

Faced with this situation, how should we as a society respond? We saw in chapter 1 that one way to determine whether values have a role to play in a particular aspect of scientific practice is by determining whether values can help scientists to achieve legitimate goals. When choosing research topics, a legitimate goal for scientists is to pursue projects that address important social priorities. Therefore, values are relevant to this aspect of science because they help to identify social priorities. For example, to determine whether research into differences in cognitive abilities should be emphasized, we need to consider whether it will help us to achieve what we value as a society. Given that one of our fundamental social values is to provide equal opportunities for everyone who has the capability to excel in science and mathematics—and given that there are strong ethical reasons to promote this value as well—it turns out that this area of research should probably not be a priority.

As philosopher Philip Kitcher has argued, research into cognitive differences is almost bound to be misused or misinterpreted in ways that lessen

2. Rose 2009, 788.

opportunities for underrepresented groups. Kitcher notes that in a culture where there are not only explicit but also implicit biases against particular gender or racial groups, research into cognitive differences is likely to be interpreted asymmetrically. On one hand, if researchers found evidence against differences in cognitive abilities, it is unclear that this would have a significant impact on society. Social scientists point out that when dealing with highly politicized issues, it is very difficult to use scientific evidence to change people's minds. Therefore, as long as there were at least some researchers still providing evidence in favor of differences in cognitive abilities, it is likely that this research would continue to have an impact in a society that already harbors social biases. On the other hand, if researchers found apparent evidence for cognitive differences between particular social groups, there is significant danger that this evidence would be taken up and applied in an unjustified manner to promote discrimination.

Suppose, for example, that researchers found small cognitive differences on average between men and women in particular skill areas, such as science or mathematics. These differences would still clearly be dwarfed by within-gender differences in these abilities. In other words, the differences in abilities between different women and between different men would be vastly more significant than any small average difference between the two genders. In recent years, it has become obvious that women have the ability to perform at the very highest levels in all areas of science. Nevertheless, women have traditionally been excluded from participation in many of these areas. Based in part on this history of exclusion, many accomplished women tell stories of how they were repeatedly discouraged from pursuing career paths in science and engineering, and they experienced discrimination in both overt and subtle ways. Moreover, there is evidence that historically disadvantaged groups, including women, internalize stereotypes about their abilities and end up underperforming academically for subconscious reasons. Therefore, any evidence for small average differences in cognitive abilities would likely end up inappropriately discouraging countless women from pursuing advanced work in math and science, despite the fact that most women who actually wanted to pursue these fields would be entirely capable of doing so.

With this in mind, why would we want to prioritize this line of investigation? No matter what it uncovered, it would be unlikely to help us achieve one of our central ethical values—namely, equal opportunity. If it appeared to provide evidence against differences in cognitive abilities, it would probably have limited impact, but if it appeared to provide evidence for cognitive differences, it would probably be misinterpreted in ways that prevented many talented people from achieving their full potential. Either way, it seems like an unhelpful addition to the sordid history of research in this area that Kourany has detailed.

One might object, however, that our society has a more diverse array of values than those that have been emphasized here. Perhaps some people value keeping women and members of other traditionally disadvantaged groups from gaining further social opportunities. But this is likely to be a fairly atypical value, and there are good ethical reasons for arguing against it. The vast majority of people across the political spectrum want to promote at least some form of equality of opportunity for everyone, and it is very difficult to justify unequal opportunity from an ethical perspective. Unfortunately, we have seen that the historical biases in our society make research into cognitive differences more likely to inhibit equal opportunity than to promote it. Therefore, given the limited resources available for supporting scientific research, this topic hardly seems like it should be a significant social priority.

Perhaps a more significant objection would be that a lack of enthusiasm for research into cognitive differences is a breach of academic freedom. One of the central values of modern universities is that scholars should be allowed to pursue whatever topics they deem to be significant, even if they are politically unpopular. In fact, one might worry that we are likely to shoot ourselves in the foot by trying to suppress research that is not "politically correct." What better way to arouse suspicions that women or other disadvantaged groups are cognitively inferior than to try to suppress research on their abilities?

Kitcher has pointed out that these objections are less convincing than they initially appear. There is a crucial distinction between arguing that research into cognitive differences should be a low priority—which is the suggestion of this chapter—versus arguing that it should be banned or suppressed. Perhaps some areas of research should actually be banned, but a variety of lesser steps could be taken in response to research on differences in cognitive abilities. For example, biologists and psychologists could voluntarily decide that they can do more good for society by pursuing alternative research topics. Also, reviewers of grant proposals are often instructed by federal funding agencies to take the "broader impacts" or the "significance" of proposed research projects into account when deciding which research projects to prioritize for funding. Therefore, it seems entirely legitimate for these reviewers to take into account the concerns discussed in this chapter when deciding how enthusiastically to review proposals for research on differences in cognitive abilities.

Perhaps another, more significant worry about the notion that cognitive differences research should be a low priority is that we may have overemphasized the extent to which this research will actually harm society. For example, despite the concerns recounted here, some scientists will surely continue to engage in this sort of research and share any cognitive differences that they find. Given the potential for this research to be misinterpreted and exaggerated, perhaps it is valuable to have other scientists who are skeptical of alleged cognitive differences but who still pursue research on this topic. That way they are available to rebut unwarranted claims, to point out weaknesses in other scholars' methodologies, and to present research that challenges unjustified claims about cognitive differences. One might also think that there are intellectual reasons for continuing to encourage research on differences in cognitive abilities. Even if it tends to be somewhat socially harmful, perhaps it is worth supporting some of this research simply because it answers questions that humans have been curious about for centuries.

These concerns provide support for the main theme of this book, which is that we need to encourage more sophisticated discussions among scientists and citizens about the roles of values in scientific research. There is room for thoughtful disagreement about the pros and cons of encouraging further research on differences in cognitive abilities, and we need to promote more reflection about these matters. As a society, we do indeed value efforts to inform our curiosity, but we need to weigh this value against our values of preventing harm to others and promoting equal opportunities for all members of society. We also have to consider whether some forms of research on cognitive differences could actually support the value of equal opportunity because they might bring clarity to debates about these issues and help train researchers who could challenge those who perform methodologically poor research and draw unjustified conclusions that promote discrimination.

These are important, value-laden issues that become relevant in a number of other areas of research as well. For example, some of the scientists who assisted the US government in the development of the atomic bomb during World War II later concluded that research on nuclear weapons was socially irresponsible. In recent years, scientific organizations have debated whether it is appropriate to publish information about deadly viruses that could potentially be used by terrorists to develop biological weapons. Research on the cultural practices or genetic characteristics of indigenous groups has also been debated because of its potential to cause harm to these already disadvantaged groups. All these examples raise difficult questions about whether it is possible to perform "responsible" research in these areas, whether our curiosity about these topics justifies research that could potentially have harmful effects, and whether it is possible to develop compromises that allow research to move forward while minimizing potential harms.

PUBLIC FUNDING OF RESEARCH: CONGRESS AND THE NATIONAL SCIENCE FOUNDATION

So far, we have seen that values can help us to identify research projects that should be a low social priority. Extending this point, we can see that values have an important role to play in deciding how to allocate public funding for research. While this is an important role for values, it raises some complicated questions that are worth exploring further. Consider a conflict that erupted in the spring of 2014 between the leadership of the National Science Foundation (NSF) and US Representative Lamar Smith of the 21st congressional district in Texas. This conflict, and the broader history of debates over the NSF, provides an excellent opportunity to explore how we can best incorporate our values in science funding decisions. In his role as Chairman of the Science, Space, and Technology Committee of the US House of Representatives, Smith asked for detailed information about grants that were funded by the NSF and that he regarded with suspicion. The leadership of the NSF resisted his request because scientists involved in reviewing each other's grant proposals expect the process to remain confidential. Ultimately, the director of the NSF, France Córdova, arrived at a compromise with Smith in which congressional staffers were allowed to come to the NSF and look over materials related to the grants without being able to photocopy them or see the names of the reviewers.

The result of this compromise was a somewhat comic scenario described by Jeffrey Mervis in the journal *Science*:

> In a spare room on the top floor of the National Science Foundation (NSF), two congressional staffers spent hours poring over confidential material relating to 20 research projects that NSF has funded over the past decade. . . . The Republican aide was looking for anything that Representative Lamar Smith (R-TX) . . . could use to demonstrate how the $7 billion research agency is wasting taxpayer dollars on frivolous or low-priority projects, particularly in the social sciences. The Democratic staff member wanted to make sure that her boss, Representative Eddie Bernice Johnson (D-TX), the panel's senior Democrat, knew enough about each grant to rebut any criticism that Smith might levy.[3]

This case illustrates the complex issues that arise when deciding how to allocate taxpayer dollars toward research funding.

The background behind this conflict is that the US government provides a great deal of funding for university scientists through various federal agencies. After World War II, when scientific research on topics like nuclear weapons and radar helped propel the US to victory, legislators and policymakers became increasingly convinced that it was valuable to pour federal money into academic research. Building on a model that was employed during the war, much of this research was funded through grants administered by agencies like the NSF, the National Institutes of Health (NIH), the Department of Defense (DOD), the Department of Energy (DOE), and the Department of Agriculture (USDA). While different agencies have somewhat different approaches, the

3. Mervis 2014, 152.

centerpiece of this funding structure is that scientists maintain a degree of autonomy from those providing the funding. The agency provides a body of money for research in a particular area, but scientists submit proposals for the research projects that they think are most promising. Then, their fellow scientists engage in a "peer review" process to look over the proposals and choose the ones that they think are best.

From the beginning, this approach raised tensions between those who wanted to promote greater freedom for scientists to choose research projects based on their own interests and those who thought that the broader values of society should play a greater role in choosing projects. Even the creation of the NSF, which was designed to focus especially on basic areas of research, got bogged down in debates over the proper role for social considerations in the funding process. In a famous essay called *The Endless Frontier*, Vannevar Bush, who led the research effort during World War II, argued that society would ultimately be best served if scientists were given maximal freedom to pursue basic research projects that were not directly tied to specific social outcomes. But Harley Kilgore, a Senator from West Virginia, argued that the purpose of federal funding for science was not "building up theoretical science just to build it up. The purpose is what has been the purpose of scientific research all the war through, and what the incentive for scientists is, to do something for the betterment of humanity."[4] Kilgore argued for greater political control over the agency and more equal distribution of funds throughout the country.

These fundamental disputes have never completely dissipated. In some instances over the past fifty years, legislators passed bills specifying how the money allocated to the NSF should be divided up among the foundation's "directorates," thereby determining how much money should go to topics like the biological sciences or the social sciences. In 1975, the House of Representatives passed legislation that would have allowed the House or Senate to reject specific grant proposals submitted to the NSF. And while these efforts to control the NSF have frequently come from conservative Republicans, it is noteworthy that Democratic President Harry Truman vetoed a bill in 1947 that he thought would give the NSF too much independence from public control. In the 1970s, another Democrat, Senator William Proxmire from Wisconsin, developed a facetious "Golden Fleece" award that he used to challenge NSF grants that appeared to be a waste of taxpayer money.

More recently, legislators such as Senator Tom Coburn (R-OK), Representative Eric Cantor (R-VA), and Representative Jeff Flake (R-AZ) attempted to limit funding from the NSF for political science projects. Following up on their efforts, Representative Smith encouraged several initiatives to

4. Quoted in Greenberg 1968, 103.

constrain the NSF. In addition to scrutinizing grants that he regarded as potentially wasteful, he proposed significantly cutting funding for social science research so that more could be spent in areas like math and engineering. He also suggested that the NSF should be required to show that each grant it funds is in the "national interest." Predictably, scientists have reacted with a great deal of frustration to Smith's efforts. In an article for the *Chronicle of Higher Education*, Glenn Gordon Smith, a professor at the University of South Florida, claimed that Smith's scrutiny of specific grants "is an outrageous politicization of science." Robert M. Rosenswig, professor at SUNY's University at Albany, was quoted as saying, "It saddens me that elected officials are attacking science in this way." Mont Hubbard, emeritus professor at the University of California, Davis, said, "Although I respect the oversight right and responsibility of Congress, I find it disturbing that the committee apparently think they can do a better job of deciding what is in the nation's interest scientifically than NSF can."

Lamar Smith responded to the skepticism from these scientists by insisting, "We all believe in academic freedom for scientists, but federal research agencies have an obligation to explain to American taxpayers why their money is being used on such research instead of on higher priorities." On one hand, it does seem reasonable for taxpayers to insist that the research they fund goes toward projects that serve their priorities. But, on the other hand, some of Smith's strategies for keeping the NSF accountable to taxpayers seem dubious at best. And some commentators have worried that Smith and many of his congressional colleagues were largely motivated by the desire to suppress research on topics like climate change that they disliked for political reasons.

In order to analyze this case more carefully, it is helpful to carve it up into three specific questions. First, should our ethical and social values influence choices about public funding for research? Second, should Congress play a central role in deciding which values are most important to the public and translating those values into decisions about how to allocate public funds for research? Third, should Congress be "double checking" the peer-review process at government agencies like the NSF to make sure that grants fulfill the values that we as a society have set for research?

Question One

Turning to the first question, we have already seen that values have a legitimate role to play in choosing the topics that researchers pursue. According to chapter 1, one way to determine if values have an appropriate role to play in a particular aspect of science is to see if they help us to achieve legitimate goals. It seems clear that at least one of our goals in public funding for scientific research is to find better solutions to problems that society cares about. Therefore, we need to take our social values into account so that we can figure out which problems are most significant to us and therefore which lines of research should be most heavily funded. Of course, this does not mean that our social values are the only considerations that we need to factor into funding choices. We may also want to consider what projects scientists think will be most helpful for advancing their disciplines or will give us the greatest insight into the natural world. Moreover, given the diversity of values in our society, it will often be very challenging to decide how to handle conflicting priorities. Nevertheless, our values are still clearly relevant to these decisions about what to fund.

Question Two

The situation gets somewhat more complicated when we turn to our second question—whether Congress should play a central role in deciding which values are important and then determining how to fund research accordingly. On first glance, it seems obvious that our congressional representatives should decide how to spend our money. But a few different worries arise. First, because our representatives have so many constituents and because they have to raise a great deal of money in order to be elected, their decisions may be more likely to track the interests of a few wealthy individuals rather than the majority of their constituents. Second, most of our representatives are not scientific experts, so it is not clear that they are particularly well equipped to decide which research projects are most likely to help us accomplish what we value as a society. Third, Congress can exert different degrees of control over the research enterprise, so even if we conclude that they should have some influence, we still have to decide what level of influence is appropriate.

Consider some of the different types of control that Congress could exert over science funding decisions. It is probably reasonable for Congress to make large-scale decisions about how much money we should spend on research for the purposes of national defense versus the amount to be spent on medical research, or the amount for agriculture, energy, or education. Even at this level, it is likely that a few powerful interest groups will skew these decisions in ways that do not serve the interests of the vast majority of the public. Unless we make significant changes to our political system, however, we have few alternatives to letting Congress play this role. Thus, given how our government operates, it seems reasonable for Congress to decide how much research money to allocate to the DOD or USDA or DOE or the NIH.

But when it comes to allocating money within these agencies, it is much less clear that Congress should be involved. It is at this point that our representatives' lack of scientific expertise becomes especially worrisome. Consider the congressional Republicans' recent antagonism toward funding political science, for example. Two political scientists from the University of Miami examined the factors that influenced how senators voted on this issue, and they found that senators with strong political science departments at universities in their states and/or who had earned bachelor's degrees themselves in political science were more likely to support continued funding in this area. This favorable attitude toward funding might just reflect a desire to support one's home-state universities and one's personal academic interests, but it could also reflect the fact that those with more understanding of political science are better able to appreciate its contribution to the social good.

One might raise similar worries about the apparent desire of Representative Smith and other legislators to challenge NSF research on climate change. The overwhelming majority of the scientific community has concluded that climate change is occurring, that human emissions of greenhouse gases are a significant contributor to this warming, and that it is likely to have significant effects on society over the coming decades. However, many members of Congress have expressed significant skepticism about this evidence. If we assume that the vast majority of the scientific community is more likely to be correct about the status of the evidence on climate change than politicians are, this raises serious questions about our legislators' ability to make decisions about climate research that promote the needs and values of society. The fear is that if our legislators challenge research funding or grant proposals related to this area of research, they are being influenced by incorrect views about the status of climate science rather than legitimate disagreements about social priorities and values. As Glenn Gordon Smith noted in the *Chronicle* article discussed earlier: "When you are selectively in denial of overwhelming scientific evidence, you seek out ways to discredit investigators who research in that area."

But perhaps we should not be too quick to dismiss our congressional representatives. Even if they are misinformed about climate science, they might be correct that the scientific community is limited in its ability to decide on its own what research projects will best serve social interests. For example, in the case of climate change, science-policy expert Dan Sarewitz has suggested that the scientific community has a tendency to focus too much attention on efforts to develop detailed models of future climate impacts, whereas funding more social-science research on strategies for adapting to climate change might better serve our social needs.

Therefore, we are left with a difficult situation. The scientific community may not always have its pulse on the values that citizens would like to prioritize, but our elected representatives may not always know enough about science to make informed decisions about which research projects will best serve our social values. We obviously need to find creative ways to integrate information about society's values with cutting-edge scientific knowledge, and there are probably better strategies than having legislators make detailed decisions about how to allocate research funding. In chapter 7, we will return to this issue and explore some additional ways of integrating citizens' values into public research funding decisions.

In sum, the answer to our second question (whether Congress should play a central role in identifying public values and allocating research funds on the basis of those values) is ambiguous. Congress should probably play some role in dividing up funds for research, but it is doubtful that they should handle the details.

Question Three

We can now turn briefly to our third question about public funding of research: Should Congress be double-checking grant proposals at the NSF to ensure that they are good investments of public funds? This seems unwise. For one thing, it opens the door for legislators to start trying to score political points by challenging grant proposals for short-sighted, simplistic reasons. It is easy for legislators on both sides of the political aisle to ridicule research projects that sound overly technical or impractical, even though they may advance our knowledge in fundamental ways that are valuable over the long term. It is also highly unlikely that legislators have the detailed expertise needed to decide whether particular grant proposals are better than others. If they are concerned that the scientific community is out of touch with significant social values, it probably makes more sense to initiate a deliberative effort of some kind that brings together a range of stakeholders to discuss their concerns with scientists and NSF officials.

Finally, it is worth noting that the NSF has specific characteristics that make it an especially poor agency for Congress to be micro-managing. While the goals of the NSF have been debated since its founding (as the conflict between Bush and Kilgore illustrates), it has always been regarded as a source of research for addressing fundamental questions that may not have immediate applications. This does not mean that the NSF must focus solely on the questions that are of the most theoretical interest to the scientific community, but it does mean that legislators should be cautious about challenging the agency's funding decisions. Our society has a wide array of values, and one of those values is the search for greater insights about the nature of the world around us. It is important to keep in mind that different federal agencies promote research that serves different social priorities, and given the mission of the NSF, it seems unwise to be overly aggressive in second-guessing scientists in their pursuit of problems that offer the most theoretical promise. Thus, we have discovered in this section that while our social values are indeed relevant to deciding how to spend public money on scientific research, there are good reasons to explore other avenues for bringing these values to light rather than allowing congressional representatives to play a detailed role in the funding process.

The previous section of this chapter focused on the role of values in deciding how the government should influence research funding. But the government is not the only or even the primary source of funding for scientific research and development (R&D). In the 1960s, the federal government supported about two-thirds of R&D in the United States, but in the twenty-first century that figure has dropped to about one-third. So if we want to think about the role of values in choosing research topics, we need to examine privately funded science as well. With that in mind, this section shows how our values are important for evaluating private-sector investments, and it uses contemporary biomedical research as an example.

As a global society, we currently face striking disparities in the global disease burden. The citizens of low-income countries deal with a number of diseases that are virtually nonexistent in wealthier countries. Consider malaria, for example. It is caused by the *Plasmodium* parasite and carried by multiple strains of the *Anopheles* mosquito, and it has been a global scourge for thousands of years. The *Plasmodium* parasite feeds on human hemoglobin, generating severe fever, chills, horribly enlarged spleens, and in some cases death. In 2013, the World Health Organization estimated that there were roughly 200 million cases of malaria and almost 600,000 deaths. To put that figure in perspective, there were almost 3,000 victims of the September 11, 2001, terrorist attacks in the United States. Thus, every couple days malaria takes as many lives globally as the 9/11 attacks. But many people in North America and Western Europe are not even aware that malaria is still a significant problem, given that it has been virtually eliminated in those geographic areas.

The absence of malaria from most wealthy countries is a relatively recent phenomenon. Science writer Sonia Shah reports that in previous centuries, the mortality rate from malaria in marshy counties near London was comparable to the rates in contemporary sub-Saharan Africa. Malaria afflicted the people of Italy for centuries, perhaps even contributing to the fall of the Roman Empire. Four popes in the fifteenth and sixteenth centuries fell prey to the disease, and in 1847 Florence Nightingale described the Campania region of Italy as "The Valley of the Shadow of Death" because of the prevalence of malaria. According to Shah, women were losing multiple husbands to the disease in the Campania region even in the first half of the twentieth century. In the United States, malaria contributed to the early demographics of the country, insofar as settlement of the southern states by whites was inhibited by the severity of malaria epidemics (whereas most people of African descent had resistance to malaria's most virulent strain). Even in the American Midwest, malaria took a serious toll on the settlers along the Great Lakes and Mississippi River. A song from the 1800s advised, "Don't go to Michigan, that land of ills, the word means ague, fever and chills."[5] During the Civil War, half of the Union troops suffered malaria at some point, including every federal soldier active in Louisiana or Alabama during 1864.

Because of demographic, agricultural, and economic trends, malaria virtually disappeared from most wealthy countries over the course of the twentieth century. In the American Midwest, for example, new practices for draining wetlands eliminated a great deal of mosquito habitat. Better infrastructure and sanitation systems also helped to eliminate breeding grounds for mosquitos. With the advent of the railroad, people were less dependent on waterways for transportation and were able to move away from swampy, malaria-prone areas. Engineers learned to design irrigation and hydropower systems in ways that did not provide breeding opportunities for malarial mosquitoes, and increased numbers of livestock in the United States and Britain meant that *Anopheles* mosquitos were more likely to bite livestock than humans.

The flip side of these successes in wealthy nations is that it is now more difficult to attract attention to the plight of those who continue to suffer from malaria in lower-income countries. After World War II, the advent of the insecticide DDT and the anti-malarial drug chloroquine created great hopes that malaria could be defeated once and for all. But soon it became clear that practical difficulties and resistance of the mosquitos and parasites to new pesticides and drugs made this a much more difficult task than it first appeared. While a variety of organizations continue to battle the disease, it still causes immense suffering around the world.

Evaluating Problematic Research Investments

The story of malaria is significant because it illustrates the worry that our current investments in biomedical research may not be directed very effectively toward alleviating the suffering of low-income people around the world. Admittedly, advances in infrastructure and sanitation may be at least as important as scientific research for addressing diseases like malaria. Nevertheless, targeted biomedical research could still play a valuable role in improving the current situation of those afflicted with malaria and a wide variety of other tropical diseases. Unfortunately, private corporations put comparatively little funding toward addressing diseases that affect the citizens of low-income nations. *Focus on the rich. (first world problems)*

Philosophers Julian Reiss and Philip Kitcher have explored this problem in detail. They point out that in the early years of the twenty-first century, diseases like malaria, tuberculosis, pneumonia, and diarrhea accounted for

CON

5. Shah 2010, 177.

about 20% of the world's disease burden (i.e., the mortality or other forms of suffering associated with diseases). Nevertheless, these diseases received less than half of 1% of all biomedical research funds. Reiss and Kitcher also point to evidence indicating that if one compares tropical and non-tropical diseases that have comparable disease burdens, the non-tropical diseases are about five times more likely to be addressed in published research articles.

What is responsible for these dramatic differences in research effort given to various diseases? The fundamental problem is that ailments like malaria and pneumonia and diarrhea and many tropical diseases disproportionately impact the citizens of low-income countries. Unfortunately, it is very difficult to motivate private companies to invest heavily in studying these sorts of diseases. It costs a great deal of money—several hundred million dollars—to bring a new pharmaceutical drug to market, and private companies obviously have to consider whether they will receive a positive return on their investments. Even if a large number of people suffer from a disease, it is difficult to receive a positive return if those who suffer from it can pay only a small amount for their healthcare. Therefore, the private marketplace is likely to put vastly more research into problems that afflict wealthy countries—even if those problems are relatively insignificant issues like impotence or heartburn or baldness—as compared to problems that primarily afflict low-income countries—even if those problems cause terrible suffering and death.

Reiss and Kitcher point out that this pattern of research funding does not accord with the ethical values that most of us espouse. They suggest that we should adopt what they call the "fair-share" principle: "at least insofar as disease problems are seen as comparably tractable, the proportions of global resources assigned to different diseases should agree with the ratios of human suffering associated with those diseases."[6] In other words, if two diseases seem to be equally easy to study and solve, then if one disease causes twice as much suffering as another, we should devote twice as many resources to addressing that disease.

Given how little research money is currently spent on the diseases that afflict low-income nations, the fair-share principle calls for dramatically shifting biomedical research toward the needs of the poor. Nevertheless, it is not entirely obvious that the fair-share principle is in fact the best ethical guideline. One might object, for example, that we surely have more ethical responsibility to address the needs of those who are close to us than those who are far away. Or one might insist that, even though it would be better if more money could be spent on the diseases of the poor, it would be inappropriate to spend money taken from the taxpayers of wealthy countries on diseases that primarily affect people in other countries. Taking this view, one could encourage

6. Reiss and Kitcher 2009, 263.

philanthropists like Bill Gates to fund research on the needs of developing countries, but one would not be able to martial the resources of the US government on their behalf. A final objection might be that even if the fair-share principle is correct in theory, it is simply too demanding to motivate meaningful change. Perhaps the fair-share principle calls for such a dramatic shift in research priorities that it is more likely to cause people to give up on acting morally than it is to convince people to promote their ethical values.

Given all these potential worries with the fair-share principle, Reiss and Kitcher wisely challenge our current system for funding biomedical research not only for its failure to address the needs of people in low-income countries but also because of the ways it fails those in wealthy countries. From a practical perspective, it is probably much easier to improve the medical situation of those in low-income countries by promoting changes that benefit the people in wealthy countries at the same time. Reiss and Kitcher's fundamental argument is that our current system for funding medical research not only harms the poor but also results in overly expensive healthcare for the citizens of wealthy countries. They argue that a large part of the problem is the patent system that gives pharmaceutical companies monopolies over the new drugs that they develop. According to Reiss and Kitcher, our current patent system raises the prices of drugs for the citizens of both poor and wealthy countries, and it does not incentivize the most beneficial avenues for medical research.

Someone who holds a patent over an invention has the right to exclude others from making or using the invention for a specified period of time. One of the most common justifications for providing patents is because they promote innovation; inventors will work hard to develop new technologies if they have the ability to patent them and require others to pay in order to use the inventions. Nevertheless, Reiss and Kitcher contend that the current patent system is far from optimal. They note that because patents give pharmaceutical companies monopoly control over their patented drugs, they generate gigantic profit margins for those companies. In the 1990s, for example, the pharmaceutical industry was making profits of about 25%, whereas other industries made about 5% in profits. Pharmaceutical profits have fallen somewhat in recent years, but they still hover between 15% and 20%, among the highest of any sector of the economy.

These sorts of profits would be more tolerable if they spurred massive levels of innovation, but it does not seem that they do so. As Reiss and Kitcher note, roughly 10% to 15% of the money generated from pharmaceutical sales goes toward research and development, whereas about 30% to 40% is spent on marketing. Moreover, they note that the patent system does not appear to provide significant incentives for companies to develop highly innovative drugs. Even when pharmaceutical companies patent new drugs, they are often just "me-too" drugs designed to mimic the action of similar drugs that are already on the market. This happens because it is much more expensive and

PRO

risky for companies to try to develop dramatically new drugs than to imitate existing drugs. Thus, the citizens of both poor and wealthy countries end up paying relatively high prices for drugs, and those drugs are not as innovative as the wealthy would prefer, nor as relevant as the poor would like.

Summarizing these points made by Reiss and Kitcher, we can see that our current patent system appears to encourage a system of pharmaceutical research that fails to meet many people's values. First, people who have little money have little influence over the decisions made by private companies. Thus, their values are given little consideration. Second, people with more money pay high prices for drugs that are often not particularly innovative. This result might seem counterintuitive, given that patents are supposed to spur innovation and research. However, the effects of patents on research depend a great deal on the details of patent policies. For example, if products are allowed to be patented even when they are not particularly novel, if patents last for overly long periods of time, and if other scientists are not allowed to perform research on patented products, then patents can cost consumers a great deal of money without pushing innovation forward very effectively. Thus, it is important to consider how policies could be altered in order to align medical research more effectively with the values of most citizens.

Potential Solutions to Problematic Research Investments

When privately funded research is unresponsive to important social values, it is not always easy to figure out how to make it more responsive. In general, the goal is to alter the incentives that companies encounter so that they shift their research priorities, but this can be difficult to achieve. For example, it is not clear how best to encourage companies to put more effort into studying the medical needs of the poor. Philosopher Thomas Pogge has suggested that the wealthy countries of the world could set aside money to put into a special Health Impact Fund (HIF). The money in the HIF could then be disbursed to pharmaceutical companies as a reward for doing research that addresses the needs of the poor. But it is difficult to create a new system like this and to get it funded. Another philosopher, James Robert Brown, has suggested that we should just abandon private funding for medical research. He argues that we could save money and produce innovations that are more relevant to social needs if we eliminated patents and depended solely on the government to fund medical research. Nevertheless, his proposal is unlikely to be politically feasible.

Even without abandoning patents, it might be possible to tweak the patent system in ways that incentivize private companies to serve social values more effectively. For an invention to be patentable under current patent law in the

United States, it must be novel, useful, non-obvious, and in accordance with other details of the statutes passed by Congress. One could alter the patent system by shifting the guidance given to patent examiners regarding the standards needed to show that an invention meets these requirements. For example, by tweaking the definition of what makes drugs "non-obvious," one could make it more difficult for companies to patent "me-too" drugs that imitate patented drugs already on the market. One could also change the statutory requirements so that companies could patent "processes" but not "products." This would enable companies to maintain their monopoly control over a particular method for producing a drug, but other companies could compete to develop alternative ways of producing the drug, thereby lowering prices. One could also lower prices by shortening the length of time that companies are allowed to exclude their competitors from using their patented products. Another possibility would be to provide greater flexibility for others to perform research on patented products as long as they do not attempt to make money off the research during the period of the patent.

Of course, even relatively minor changes will probably be very difficult to enact. Pharmaceutical companies make enormous profits, and they are likely to be very resistant to changes to the patent system that would affect their bottom line. However, it might be possible to find ways of tweaking the patent system or providing alternative incentives that would be palatable to the pharmaceutical industry while shifting medical research in socially desirable directions. One of the important lessons to draw from this section is that it is not enough for us to focus solely on the values of individual scientists or even groups of scientists when thinking about how values influence the choice of research topics. Institutional policies, laws, and regulations can steer scientific research in directions that serve some values and detract from others. Therefore, we need to address these institutional factors when attempting to alter how values play a role in scientific research. Chapter 7 will return to these questions about how to guide the values that influence scientific research and that are supported by it.

CONCLUSION

Whereas the later chapters of this book explore somewhat more controversial avenues through which values play a role in science, almost everyone can agree that values have a legitimate role to play in choosing research topics. Nevertheless, a number of issues arise when one explores the specific ways in which values influence this aspect of science. This chapter considered three of these issues. First, it examined how values can play a role in making difficult decisions about which research projects to prioritize over others. Second, it explored questions about how our values should be incorporated into public research funding. Third, it examined how values can play a role in evaluating and influencing private-sector research.

First, we saw that some areas of research, like studies of cognitive differences or weapons, have the potential to generate harm for specific groups or for society as a whole. When they seem likely to cause harm, it is reasonable to accord these research projects low priority compared to other research topics that could be more beneficial for society. Nevertheless, when deciding whether or not to pursue potentially harmful lines of research, we need to consider the benefits of having well-intentioned people with research experience in these areas so that they can counteract those with more pernicious intentions.

Second, we saw that it can be difficult to figure out how best to steer publicly funded science so that it accords with social values. Legislators clearly have a significant role to play in deciding how taxpayer money should be spent, but they do not appear to be well-qualified to make detailed decisions about how to fund scientific research. Especially in the case of an agency like the NSF, which is designed to support fundamental areas of research, there is much to be gained by giving the scientific peer review process a great deal of latitude. But it is still important to consider how information about pressing social needs and priorities can best be incorporated into funding decisions, and this is a question to which we will return later in the book.

Third, we considered how social values can be used to critique the funding priorities of the private sector. In theory, market forces should incline private corporations toward studying what people value, but this process is imperfect. For example, it can fail when people are too poor to influence the market or when patent policies create monopolies that eliminate competition. One of the significant features of our analysis of biomedical research is that it illustrates that value influences can be embedded in institutions and social systems, not just in the minds of individual scientists. For example, the patent system creates incentives that tend to steer research in particular directions (e.g., the development of drugs for ailments that afflict the wealthy). Therefore, even if individual scientists—or individual executives within pharmaceutical companies—have deep social concern for the needs of those with low incomes, the forces at work in the marketplace severely limit their ability to act on those values. Thus, large-scale legal and policy changes may be needed in this and other cases in order to promote the values that we care about within the scientific enterprise.

SOURCES

For more information about Larry Summers's biography, see Bombardieri (2005, 2006), Ciarelli and Troianovski (2006), and Plotz (2001). Kitcher (2001) and Kourany (2010) provide excellent analyses of the roles of values in research on differences in cognitive abilities.

To learn more about the recent and historical wrangling between the NSF and Congress, see Kintisch (2014). Basken (2014) wrote the *Chronicle of Higher Education* article with quotations from scientists responding to Lamar Smith's proposals. Greenberg (1968) provides an excellent summary of the debates before and after World War II about the extent to which scientists should be given freedom to choose their own research projects. Bush (1945) is a classic statement of the view that giving scientists freedom to do basic research ultimately serves society. Uscinski and Klofstad (2010) analyzed the factors that influenced senators' votes on political science funding.

Shah (2010) provides an excellent overview of malaria and its impact on humanity. Reiss and Kitcher (2009) examine the questionable allocation of resources for biomedical research in both developing and developed countries. Pogge (2009) contains an overview of his vision for the Health Impact Fund, and Brown (2002) provides his argument for socializing medical research. Biddle (2014b) provides a very accessible overview of the patent system and the ways it could be altered to promote more socially beneficial innovation.

CHAPTER 3
How Should We Study It?

In the year 2000, a landmark paper in the journal *Science* reported on the successful genetic modification of a strain of rice so that it produced beta-carotene, which is converted by the human body into vitamin A. This was regarded as a hugely significant breakthrough because millions of poor people around the world suffer from vitamin A deficiency, resulting in hundreds of thousands of deaths and cases of blindness every year. Rice is a staple food for many of these people, so if it could be modified to produce beta-carotene in the edible portion of the plant, it could potentially alleviate a great deal of suffering. The plant was dubbed "golden rice" because the beta-carotene gave it a yellowish hue. A humanitarian project like this hardly seems to be the sort of research that would arouse significant controversy, but it became a flashpoint for major global debates.

As we will see in this chapter, the debates over golden rice illustrate major rifts over the best strategies for moving contemporary agricultural research forward. Many researchers think that the most promising options involve the genetic modification of crops so that they can resist herbicides, grow well in hostile conditions, and even produce their own pesticides. Others think that it is more important to study traditional, more ecologically friendly agricultural techniques, in order to find ways to improve the efficiency of these approaches. Still others think that we can use all these techniques in combination to address the world's agricultural needs. These diverse ways of pursuing agricultural research illustrate the major theme of this chapter, namely, that values have an important role to play in deciding *how* to pursue a particular research topic. We will explore how the methods that scientists use, the assumptions that they make, and the questions that they ask can all vary when studying the same topic, depending on the explicit or implicit values that influence the research.

To drive home the point of the chapter, it is illuminating to consider the story told by many Eastern religious traditions about a group of blind men who encounter an elephant. One blind man feels the elephant's trunk, another one encounters the tail, yet another one feels its side, and so on. As a result, they come to widely diverging accounts of what an elephant is like, even though they are all describing the same creature. This story is an apt analogy for the multiple ways of studying many scientific topics, including agriculture. Over the long run, scientists might ultimately converge and provide a unified account of the topic under investigation. But scientists often persist in providing competing accounts and studying different aspects of a phenomenon for very long periods of time. When the subject of an investigation is socially important, these disagreements about how to approach a research topic can have major ramifications for society and can support very different values.

As emphasized in the previous two chapters, one way to determine whether values have an appropriate role to play in a particular aspect of scientific research is to determine whether values are relevant to achieving legitimate goals. In the cases described in this chapter, a legitimate goal for scientists is to investigate their research topics in a manner that best addresses social concerns and priorities. Consider the case of agriculture. If society decides that it is important to build up the agricultural biotechnology sector, then it makes sense for scientists to approach agricultural questions in a manner that fits with this technological agenda. In contrast, if society concludes that social and environmental needs are better addressed by improving traditional agricultural approaches, then this ought to influence how scientists approach the topic.

This case also fits the other justification discussed in chapter 1 for incorporating values in science. This justification is that value influences are often inevitable; no matter how a scientific choice is made, it will end up supporting some values over others. We will see in this chapter that the choice of how to study a research topic often fits with this justification, insofar as any approach to studying a topic will end up providing more support for some values and less support for other values. If scientists will end up advancing some social goals over others no matter how they study a topic, it is best for them to recognize this fact and to reflect carefully on which values should influence their research.

This is not to say that social values are the only factors that are of importance when deciding how to study a research topic. Scientists also have to consider the scientific feasibility of different approaches and the ways in which various research strategies will promote progress in their disciplines. But social values clearly have a role to play alongside other legitimate considerations. The key is for scientists to find ways to make these value influences explicit so that people are not confused about the value-laden assumptions and methodological choices that have influenced the research. The following three sections explore three different ways in which a topic can be studied differently, depending on the values that influence the investigation: (1) different *methodologies* can be used, as we will see in research on agriculture; (2) different *assumptions* can be made, as we will see in research on industrial pollution; and (3) different *questions* can be asked, as we will see in medical research.

RESEARCH METHODS IN AGRICULTURE

One way for values to influence the study of a particular research topic is by affecting the methods that are used for investigating it. Value-based disputes over methods are evident in recent debates over the best ways to pursue agricultural research, as illustrated by the conflict over golden rice. For example, the environmental organization Greenpeace labeled golden rice "fool's gold." Greenpeace opposes the genetic modification of plants because of the concern that they could generate environmental problems, and the organization pointed out that it hardly seemed worth taking a risk on planting golden rice given that it contained far less than the daily recommended allowance of vitamin A. The environmental activist Vandana Shiva similarly called golden rice a "hoax," insisting that it would actually aggravate vitamin A deficiency rather than solving it. She claimed that "the world's top scientists suffer a more severe form of blindness than children in poor countries,"[1] because they do not recognize the wide variety of other plants that poor people can eat in order to obtain vitamin A. She complained that the development of genetically modified crops merely exacerbates the trend in low-income countries toward adopting industrial agricultural practices. In her view, these practices are the real cause of vitamin A deficiency because they encourage farmers to grow only a single crop in their fields and to kill all other plants (including greens that are high in vitamin A) with herbicides.

Proponents of golden rice have responded with outrage to these critics. Ingo Potrykus, one of the main scientists involved in developing the crop, insisted that to block the development of golden rice by destroying test fields would be to contribute to "a crime against humanity."[2] Gordon Conway, the president of the Rockefeller Foundation at the time, insisted that Vandana Shiva did not adequately consider the range of situations in which poor people

1. Shiva 2002, 60.
2. Potrykus 2002, 57.

find themselves. He argued that fruits and vegetables are often expensive or difficult to access, and therefore many people subsist primarily on rice. He also pointed out that scientists were striving to increase the levels of beta-carotene in golden rice, so it was short-sighted to criticize the rice for having too little of this nutrient during the earliest stages of research. Furthermore, he noted that even small increases in poor people's access to vitamin A could be beneficial, even if the rice did not provide all of the recommended daily value. Finally, he contended that even if Indians are too dependent on rice for their nutrition, it still makes sense to increase its nutritional value until they can switch to eating a greater variety of foods.

The critics of golden rice have not given up. They argue that this crop is culturally problematic because most people in Asia strongly prefer white rice. Apparently, asking these people to eat yellow rice would be a bit like asking Westerners to eat blue bread. The critics also argue that the beta-carotene is easily lost from the rice if it is not cooked properly, and the proper cooking methods are difficult for many poor people to accomplish. They also worry that if people do not have enough fat in their diet, the beta-carotene is of little nutritional value to them. Based on these considerations, many critics insist that golden rice is a "technical fix" that does not address the real problems that need to be addressed. They want to see social and agricultural policies changed in order to alleviate poverty rather than trying to apply technological "band aids" to the problem.

Building on the global tensions over golden rice and other crops, a group of agricultural biotechnology companies asked the World Bank and the Food and Agriculture Organization (FAO) to provide advice on the future of genetically modified crops in developing countries. Their request launched a major project to study agricultural science and technology and its role in improving the lives of people in the developing world. The project was ultimately sponsored by a range of important international organizations including the United Nations Environment Program (UNEP), the United Nations Development Program (UNDP), and the World Health Organization (WHO). Hundreds of experts were chosen to participate as authors, many governments and organizations participated in reviewing the report, and the organizers hoped that it could have a global influence comparable to that of the Intergovernmental Panel on Climate Change (IPCC). Unfortunately, its success was hampered by the fact that major agricultural companies like Monsanto and Syngenta backed out before the end of the project, and several countries (including Australia, Canada, and the United States) provided only partial support for the final documents. As one commentator concluded, "there is a sense of having lost a wonderful opportunity."[3]

3. Stokstad 2008, 1474.

How did this project—called the International Assessment of Agricultural Knowledge, Science, and Technology for Development (IAASTD)—break down? The project's difficulties cannot be isolated to a single factor, but at least one major problem is that different participants wanted to pursue distinct methods for addressing agricultural problems. Philosopher Hugh Lacey refers to these differing approaches as "research strategies," and he argues that they are highly significant because they tend to shift society in different directions. For example, Lacey argues that contemporary agricultural research is largely characterized by one particular set of methods. These methods focus on maximizing agricultural production by analyzing the genetic characteristics of crops and manipulating those genetic traits to produce high-yielding seeds. These approaches are frequently conjoined with research efforts to identify specific inputs, such as fertilizers and pesticides, that can be added to the crops to further increase production.

One of the advantages of these approaches to agricultural research is that they feed smoothly into the activities of agricultural biotechnology companies. The companies can generate significant profits by patenting new fertilizers, pesticides, and seed varieties. Recently, these companies have developed even greater synergies between their products by genetically modifying seed varieties to pair with herbicides that kill everything other than the modified seeds. Since the 1980s, a number of US government policies have been designed to encourage research universities to develop collaborative research projects with private companies like these. Finally, all these biological research efforts are often conjoined with economic analyses to determine how specific approaches to agricultural production are likely to affect national and global economies.

While some figures involved in the IAASTD report sought to advance these typical research methods, many others wanted to pursue different research strategies. For example, the IAASTD report's *Global Summary for Decision Makers* focuses on a very broad question: "How can AKST [agricultural knowledge, science, and technology] be used to reduce hunger and poverty, improve rural livelihoods, and promote equitable environmentally, socially, and economically sustainable development?"[4] To answer a question like this, it is necessary to employ a range of interdisciplinary methods that draw on insights from ecology, environmental science, and sociology. The motivation for taking these broader approaches is that when agricultural research focuses primarily on techniques that maximize production, it can ignore negative side-effects that accrue to small-scale farmers and natural resources, including forests and

---

4. IAASTD 2009, 3.

fisheries. Given that these natural resources are often central to the livelihood of farmers in developing countries, focusing solely on agricultural production tends to enhance some aspects of agriculture (specifically, the production of major agricultural commodities) while producing a net loss in quality of life for many poor community members and small-scale farmers.

Because of these concerns about poverty and the environment, critics of contemporary agricultural research argue that many economic analyses of agriculture are seriously inadequate. These analyses often focus on gross domestic product (GDP), which is a measure of the market value of all goods and services produced by a country. GDP can be very misleading, however, when used to draw inferences about the well-being of a society. It does not reflect inequities in the wealth of a population, it does not reflect the ways in which economic production may be depleting natural resources, and it does not consider whether the production of goods and services is being directed toward purposes that we would have preferred to avoid in the first place (e.g., repairing damage caused by natural disasters or accidents). Because of these worries, the IAASTD report emphasizes that "measurements of returns to investments [in AKST] require indices that give more information than GDP, and that are sensitive to environmental and equity gains."[5] Moreover, even when steps have been taken to measure the impacts of agriculture on the environment and on people's livelihoods, the report emphasizes that there is still very little information about the trade-offs and relationships between these various measures. Therefore, the report calls for new methods of evaluating agriculture that focus on these broader questions.

Alternative Research Methods

The IAASTD report suggests a variety of alternative research methodologies that might be more effective at reducing hunger and poverty while promoting environmental sustainability. One of the report's recommendations is to pursue research on agroecology, a field of agricultural science that looks for ways to grow crops in ecologically friendly ways. Agroecologists often recommend planting multiple crops in the same fields, with the goal of promoting better soil development, healthier ecosystems, and less need for added fertilizers and pesticides. They also study how agricultural animals can be integrated into farming systems so that the waste from the animals can contribute to greater crop production. These agroecological approaches sometimes require more labor and can therefore result in somewhat higher prices for agricultural products, but they can generate very large quantities of food with relatively few

5. IAASTD 2009, 7.

input costs. They can also provide further opportunities for women to become involved in agricultural production. Thus, in rural areas with numerous poor agricultural workers, agroecological strategies can sometimes promote more economically, physically, and environmentally healthy communities.

Other promising methodologies come from the social sciences. Whereas the dominant contemporary approaches to agricultural science typically focus on techniques from the natural sciences, incorporating insights from the social sciences can help to develop agricultural approaches that strengthen rural communities and alleviate rural poverty and hunger. For example, the IAASTD report suggests increasing poor farmers' security of access to land, promoting more transparent and high-functioning markets, and providing better social safety nets.

Historically, these approaches from the social sciences have often been given short shrift because the dominant social and political values favored methods from the natural sciences. As an example, the US government pushed very strongly in the 1960s to address the problem of hunger in India as a set of technical problems to be addressed with Green Revolution technologies (e.g., high-yielding seeds, fertilizers, pesticides, and irrigation). This was partly because the major alternative approaches (e.g., redistributing land to make it available to poor peasants) sounded far too much like a Red (Communist) Revolution, which was untenable during the height of the Cold War. Similarly, when the government of Guatemala made plans in the 1950s to force the sale of unused land from large plantations in order to make the land available to poor farmers (who had been historically dispossessed because of colonial policies), the US government orchestrated a coup in order to halt this land reform. Given the controversial nature of these social and political values, it is not surprising that our approaches to improving agriculture in the twentieth and twenty-first centuries have focused primarily on technical methods from the biological sciences as opposed to interventions coming from the social sciences.

It is important to keep in mind that the authors of the IAASTD report are not completely opposed to more biologically and chemically intensive research approaches. For example, they note that high-yield seeds and genetically modified crops may be valuable in some cases. Nevertheless, the report emphasizes that the tight patent protection that biotechnology companies currently maintain over their GM seeds tends to prevent local farmers from engaging in effective collaborative research with industry scientists. As a result, the report expresses skepticism about whether these seeds, which have been so popular among many large-scale farmers, will adequately meet the needs of small-scale farmers in low-income countries. These criticisms of current patent policies and GM seeds were major factors that undermined support for the IAASTD report among biotechnology companies and countries like the United States.

In sum, agricultural research illustrates the relevance of values for deciding how to go about studying a particular research topic. Many different methodologies can be used for studying a complex issue like agriculture: investigating the genetic properties of seeds, developing new pesticides, performing economic analyses of specific agricultural techniques, examining synergies between multiple crop and animal species, examining how land reform affects poor farmers, and studying how different agricultural techniques affect the health of local ecosystems. Different methods tend to serve different social values. For example, focusing on the genetic properties of seeds and developing new pesticides promotes the creation of flourishing agricultural biotechnology companies and big farming operations that yield large quantities of grain at relatively cheap prices. In contrast, focusing on the synergies between multiple crops or studying how land reform would affect poor farmers may be more likely to generate solutions that serve the needs of poor farmers in low-income countries. Deciding which values to prioritize, and therefore which methods to emphasize, is a difficult issue that requires reflecting about our ethical commitments and engaging with affected stakeholders (as discussed in chapter 7).

ASSUMPTIONS IN POLLUTION RESEARCH

Values can also alter the assumptions that scientists make over the course of their investigations. In order to perform their research successfully, scientists have to make assumptions about a wide range of issues, including what counts as legitimate evidence, which methods are best suited for collecting and analyzing evidence, how best to interpret evidence, and how best to handle gaps or limitations in evidence. This role for values is illustrated by the story of Woburn, Massachusetts, which has become famous as a site of industrial contamination. Located a few miles north of Boston, it has been the home of leather tanneries, chemical factories, and paper-producing facilities. In the 1970s, people became concerned about illnesses in the community. Especially devastating was a group of childhood leukemia cases among a handful of families who lived very close to one another. Anne Anderson, whose son Jimmy was diagnosed with leukemia when he was four years old, suspected that the local water supply might be contributing to these illnesses. The water was known to have a strange color and taste at times, but most experts and citizens doubted that it was the source of people's health problems. This began to change in 1979 when barrels of industrial chemicals were found buried near two of the city's wells. Tests revealed that the wells contained harmful pollutants, including the chemical trichloroethylene. Although the pollutants in the wells did not match the chemicals in the barrels, these findings stimulated further efforts to investigate the cause of the pollution.

These investigations ultimately led to a lawsuit against two corporations—Beatrice Foods and W. R. Grace—that owned facilities near the wells. The lawsuit became internationally famous because the writer Jonathan Harr followed the plaintiffs and their attorney, Jan Schlichtmann, throughout the case. His book, *A Civil Action*, inspired a feature film of the same name, starring John Travolta as Schlichtmann. The legal case was fascinating in part because it pitted the sophisticated lawyers of Beatrice and Grace against Schlichtmann's small law firm, which could barely handle the expense of such a long and complicated case. The presiding judge finally concluded that there was not enough evidence to proceed against Beatrice, and Grace eventually settled with the families for a reported $8 million. Even after the settlement, both Beatrice and Grace were engaged in further litigation with Schlichtmann and the EPA concerning their dumping of toxic chemicals. Thanks in part to the popularity of the book and movie, the case drew public attention both to the hazards of industrial pollution and to the very complex process of achieving justice for the victims of environmental harms.

One of the greatest challenges for families in cases like this one is the difficulty of providing decisive scientific evidence to prove that particular chemicals caused their specific illnesses. Because diseases like cancer can be caused in many different ways, it is challenging to identify the specific factors that are responsible in a particular case. Furthermore, it is very difficult to determine the range of ailments that can be caused by a particular chemical. It would be unethical to expose people to toxic chemicals in a controlled experiment, which means that scientists have to depend on animal experiments and indirect observations of chemical effects on humans. In part because of these limitations, scientists are forced to make significant assumptions when trying to draw conclusions on the basis of the available evidence.

In the Woburn case, Anne Anderson and other community members initially had little success in convincing local and state regulatory agencies to take their concerns seriously. Nevertheless, she and her local priest took the initiative to meet with affected families and to map the locations of other leukemia victims. Even though a study by the Massachusetts Department of Public Health (DPH) concluded that the evidence for an association between the drinking water and leukemia in Woburn was inconclusive, the concerned citizens questioned the study's conclusions and convinced researchers at the Harvard School of Public Health to help them perform another study. Community volunteers performed thousands of phone interviews in order to collect information about the health of Woburn residents, and an analysis by the Harvard researchers ultimately found a significant correlation between health problems and consumption of water from the contaminated wells. To describe how citizens in Woburn and other towns collaborated with scientists, sociologist Phil Brown coined the term "popular epidemiology." The Woburn case is now regarded as a precedent for many subsequent efforts to engage in "community based participatory research," which we will study further in later chapters.

One of the major benefits of incorporating citizens alongside scientific experts in research efforts is that citizens can help to challenge methodological assumptions that influence the selection and interpretation of research results. Just as the DPH obtained different results from the Harvard School of Public Health, multiple studies of health effects caused by environmental pollution frequently yield conflicting findings. Deciding how much to trust particular studies frequently depends on evaluating methodological assumptions. For example, assumptions about the appropriate boundaries of the geographical area to be studied can be very important to scrutinize; if a study incorporates some heavily polluted areas and other areas that are not very polluted, it can make pollution threats appear less serious than they would otherwise be. Similarly, analyzing health effects together for two neighboring towns might yield statistically significant evidence for health problems, whereas analyzing health effects in the two towns separately might not yield statistically significant results. Additional assumptions arise when researchers have to decide how long their studies should run and which people to include in the studies. In the Woburn case, for example, critics of the DPH study worried that it had not included some children who moved away from Woburn and some nonresidents who spent periods of time with family members in the town. If the community members had not persisted in calling for further studies of their community, however, the DPH study might never have been scrutinized more carefully.

Assumptions Highlighted by the Madison Environmental Justice Organization

For another example of the ways in which citizen groups can question expert assumptions, let us consider the work of the Madison Environmental Justice Organization (MEJO), a small multicultural community organization in Madison, Wisconsin, that focuses on alleviating unequal exposures to environmental threats. One of the organization's central goals has been to address the exposure of subsistence anglers to toxic substances in the fish that they catch and eat. State and federal regulatory agencies collect information about toxicants in fish and inform citizens about hazards via advisories. But Maria and Jim Powell, two members of MEJO, have highlighted two questionable assumptions that have influenced these agencies. First, the agencies have made the questionable assumption that Madison lakes are relatively non-toxic compared to other waterways in Wisconsin and therefore do not merit serious scrutiny. Second, they have assumed that subsistence anglers are consuming less fish—and less contaminated portions of fish—than they probably are.

Scientists may not deliberately choose assumptions like these because of the ways they support their social values, but those assumptions may nevertheless tend to favor the interests or values of some social groups over others. For example, the Powells have argued that the assumption that Madison lakes are not significantly contaminated stems from social and cultural "blinders" that prevent many researchers from paying attention to the risks faced by minority groups that consume a great deal of fish from Madison lakes. They also contend that a "chicken-and-egg" situation ensues because researchers do not bother to collect data that could challenge their initial assumption that the lakes are safe.

MEJO has also collected evidence indicating that a number of subsistence anglers eat fish (including large fish that are more polluted) every day or several times a week. Moreover, some immigrant groups eat the entire fish (sometimes with organs), which can also increase their exposure to toxic chemicals. As a result, government risk assessments that assume lower fish consumption and different sorts of fish consumption are likely to underestimate the risks faced by these groups. Therefore, this case illustrates how seemingly objective and straightforward scientific analyses can end up disadvantaging social groups that are already marginalized because the analyses include questionable assumptions. In some cases, like the one considered here, the assumptions can be shown to be false without too much difficulty. In other cases, it is very difficult (at least in the short term) to determine which assumptions are actually correct, so scientists face significant and value-laden choices about which assumptions to adopt as they perform their research.

This can happen, as it did in the MEJO case, because the experts do not understand all the behaviors of local citizens. In the MEJO case, the exposure occurred through fish consumption, but experts can also misjudge the behaviors of farmers that lead to pesticide exposures or the playing habits of children that result in exposures to hazardous substances. Sometimes citizens also disagree with the ways that experts average exposure levels across groups of people or across time. They worry that these averages can hide more acute exposures and health effects, resulting in studies that underestimate the effects of toxic substances. Of course, not all experts display these tendencies, and not all citizens have the same worries. Moreover, citizens sometimes make assumptions that are later shown to be false or misleading. But the important point is that citizen involvement in public-health research during recent decades has highlighted that studies of environmental pollution often incorporate important assumptions that merit further investigation.

Consider a few of the other assumptions that scientific experts often need to make when studying environmental pollution. First, they are often forced to make assumptions about the effects that a toxic chemical will cause at very low doses. In order to save money and generate more straightforward results, most animal experiments focus on relatively high doses of toxic substances, which means that experts have to make challenging decisions about how to model the likely effects at lower dose levels. Second, experts often have to make assumptions about whether the harmful effects observed in animal studies are likely to be observed in humans as well. This can become even more confusing when preliminary studies in humans do not provide evidence of harm. The lack of effects could be caused by limitations in the human studies, or they could indicate that humans are not affected in the same way as other animals. Third, experts sometimes have to make assumptions about whether—and to what extent—sensitive groups of people like children and pregnant women will suffer greater harmful effects from toxic substances than other people. Fourth, they are sometimes forced to make assumptions about how the effects of toxic chemicals in combination will differ from the effects of each toxic chemical considered individually. Fifth, experts sometimes disagree in their assumptions about whether new methods of research are reliable or not. When new methods indicate that a chemical may be harmful and more traditional methods indicate that it seems to be safe, experts are often faced with very difficult decisions.

Debating Whether Values Should Affect These Assumptions

Skeptical readers might question whether these choices about assumptions really constitute situations where values ought to influence scientists. Even if values do often influence the assumptions that scientists make, one might argue that scientists should try to resist these influences and avoid making assumptions until they can collect further evidence to guide their decisions. A major problem with this position is that it is unlikely that scientists could perform much of any research without holding at least some assumptions that go beyond the available evidence. For example, philosopher Helen Longino has argued that scientists always face at least some "gap" between the evidence available to them and their conclusions, and they are forced to fill this gap with background assumptions about what counts as legitimate evidence for a particular conclusion.

Furthermore, even if one denied that there were always a gap between scientific evidence and conclusions, it seems clear that scientists are often asked to inform citizens and policymakers about topics (such as the health effects of toxic chemicals) for which they have limited evidence. In many cases, it would be unreasonable for them to decline to offer any conclusions until they had such decisive evidence that they could avoid making assumptions; this could take an exceedingly long time. It would also be unreasonable for them to make assumptions without considering their social consequences. In the cases considered in this chapter, some assumptions make suspicious chemicals more likely to appear toxic, thereby promoting avoidance of the chemicals and leading to a potential improvement in public and environmental health. Other assumptions make suspicious chemicals less likely to appear toxic, thereby serving various economic interests. It would be irresponsible for scientists to ignore these consequences of their work. Thus, these are cases where value-laden choices cannot be avoided, and so it is best to make the choices transparently and thoughtfully.

Perhaps the opposition to allowing values to influence these assumptions stems from the false impression that it would harm the objectivity of science and end up rendering these choices arbitrary. On further examination, this is not the case. Recognizing the need to make value-laden assumptions can actually make science more rigorous and transparent. For example, Longino calls for the development of venues in which scientific work can be critically evaluated so that key assumptions can be uncovered, the status of the available evidence can be clarified, and scientists can explain why they are making some assumptions rather than others. These critical venues can involve traditional institutions like journals and conferences, but they can also involve more innovative approaches that incorporate interdisciplinary groups of scholars and citizens (see chapter 7).

In the case of environmental pollution, scientists have already been working with policymakers and other stakeholders to decide how to handle many of the assumptions involved in assessing risks from toxic chemicals. For example, the US EPA has developed guidelines that advise scientists about how to extrapolate from effects in animals to effects in humans, from high-dose effects to low-dose effects, and from effects in adults to effects in children. In this case, it is widely recognized that these assumptions have important social consequences, and the EPA has tried to choose them in a manner that best serves public health. Cases like this illustrate that it is probably more dangerous for scientists to think that they can avoid making value-laden assumptions (in which case they are likely to subconsciously suppress or hide them) than for scientists to acknowledge their assumptions and discuss them openly.

QUESTIONS IN MEDICAL RESEARCH

Values can also alter the questions that scientists ask about a particular research topic, as one can see in many cases involving medical research. We saw in the last chapter that the pharmaceutical business is one of the most lucrative sectors of the economy. Whereas most industries make profits of 5% or so per year, profits in the pharmaceutical industry have typically ranged between 15% and 25% over the past 25 years. We also saw that pharma has come under criticism for not being as innovative as many people would like. But this criticism is only the tip of the iceberg. A host of books have assailed the pharmaceutical industry's practices; consider just a few prominent titles: *Bad Pharma: How Drug Companies Mislead Doctors and Harm Patients, The Truth about the Drug Companies: How They Deceive Us and What to Do about It, On the Take: How Medicine's Complicity with Big Business Can Endanger Your Health*, and *Deadly Medicines & Organised Crime: How Big Pharma Has Corrupted Health Care*.

One can get a taste for the major concerns these books raise about the pharmaceutical industry by considering the stories of three top-selling antidepressant drugs: Paxil, Zoloft, and Prozac. In the late 1990s, GlaxoSmithKline, the manufacturer of Paxil, conducted a number of trials to determine its effectiveness in children. The studies indicated that the drug was no better than a placebo; even worse, they suggested that it might contribute to an increased risk of suicide. Unfortunately, GlaxoSmithKline did not make either of these findings known to the general public or to regulators until much later. At one point, they did send information about Paxil's suicide risks to the United Kingdom's regulatory agency, but they mixed studies with children together with studies in adults, so the evidence of increased suicide risk in children disappeared. While companies are legally required in some cases to disclose negative information about their products to regulators, they do not always publish this information for the general public. For example, one study of publications on antidepressants found that of 74 major research studies that were conducted by pharmaceutical companies, the results from 37 out of the 38 favorable studies were published for the public, but only three of the 36 negative studies were published.

Even when studies are published, the story of Zoloft illustrates that the publications are often carefully controlled by the manufacturer. Because of documents made available through litigation, scholars were able to access information about Pfizer's plans in the 1990s for publishing papers about its drug Zoloft. They found that Pfizer had contracted with a company called Current Medical Directions (CMD) to produce a number of journal articles about the drug. These articles were written by CMD, and then prominent academics were solicited after the fact to put their names on the papers. This process of writing scientific articles through a company like CMD and then finding academics to serve as nominal "authors" (presumably for the sake of increasing the impact of the articles) is called ghostwriting. It is difficult to determine how often it occurs because ghostwritten articles typically do not acknowledge the actual process of authorship. Nevertheless, in the Zoloft case, it turned out that more than 50% of the articles published between 1998 and 2000 were ghostwritten by CMD. These articles were almost entirely positive about Zoloft, whereas only about half of the articles written outside of Pfizer's control reported positive results. Nevertheless, the ghostwritten articles were published in more prestigious journals and were more widely cited in the subsequent literature.

Consider also the case of Prozac, which is the marketing term for the anti-depressant drug fluoxetine hydrochloride, produced by Eli Lilly. In the early 2000s, Lilly began to promote the idea that fluoxetine hydrochloride could be used as an effective treatment for an ailment called premenstrual dysphoric disorder (PMDD) to which the company was drawing a great deal of attention. This ailment is closely related to premenstrual syndrome (PMS), insofar as it is characterized by the following symptoms right before the onset of menstruation: a markedly depressed mood; marked anxiety; feeling tense, anxious, or "on edge"; feeling fatigued, lethargic, or lacking in energy; and so on. A number of commentators found it extremely suspicious that Lilly was aggressively promoting the idea that PMDD was a new disorder at around the same time that they were losing much of their patent protection on Prozac. By identifying a new disorder that the drug could treat, Lilly could gain further patent protection and prevent some of its profits from disappearing. The skeptical commentators argued that there was no legitimate reason for distinguishing PMDD from PMS and that Lilly was trying to promote a new disease concept primarily for financial reasons.

Values in Research Questions

These activities by the pharmaceutical industry are of great concern, but they are all relatively obvious. It is fairly clear that policymakers need to take steps to halt the suppression of negative data, the surreptitious ghostwriting of papers, and the creation of new disease categories for marketing purposes. The remainder of this chapter focuses on a more subtle way in which values influence the pharmaceutical industry. Specifically, they influence the questions that the companies ask when they are doing research. It might seem like this issue belongs in the previous chapter, where we explored how values can influence what researchers choose to study. Nevertheless, it is appropriate to treat this issue separately here because researchers can choose a particular topic for investigation and still ask very different sorts of questions about it. By asking different questions, they can learn about different aspects of the phenomenon and thereby promote very different values. Thus, along with choosing different methodologies and different assumptions, the choice of specific research questions is a third significant way in which values can influence how a particular topic is studied.

Let us see how this has played out in the case of research on depression. Depression is a highly complex condition that can be influenced by a number of different biological, psychological, and social factors. Therefore, efforts to study its causes and treatments can take a number of different forms, including efforts to provide individuals with coping strategies or to initiate change in their social environments. Even efforts to understand depression at a biological level can take multiple forms, including investigating how sleep or dietary patterns or exercise can affect it. But recent efforts to address depression, especially within the field of psychiatry, have focused primarily on studying how it is related to neurochemical changes in the brain and how various drugs can alter that neurochemistry. Paxil, Zoloft, and Prozac are all examples of selective serotonin reuptake inhibitors (SSRIs), which have been widely used for treating depression in recent years. In sum, much of the recent medical research on depression has focused on questions about the molecular pathways that contribute to the phenomenon and the possibility of developing drugs that could alter those pathways. Nevertheless, some scholars claim that when the available evidence is not manipulated by the pharmaceutical industry, it actually indicates that approaches like psychotherapy are more effective than SSRIs for many people.

A number of values have contributed to this research focus. For example, it appears that psychiatrists thought they could gain more scientific legitimacy for their field in the 1970s and 1980s by focusing more attention on biological explanations for mental problems rather than psychological or social explanations. Another crucial value consideration is that pharmaceutical companies can make a great deal of money marketing patented drugs as treatments for depression, whereas solutions like exercise—which appears to be very effective for many people—do not generate major financial returns for those who investigate them. Thus, medical research has focused on asking questions about a narrow range of biological phenomena that can be manipulated by patentable, financially lucrative interventions.

Contemporary research on cancer provides another example that vividly illustrates how medical researchers can ask different questions about a particular topic depending on the values that drive the research. Much cancer research has focused heavily on treatment, including radiation and chemotherapy in particular. Some individuals have worried that comparatively little money has gone toward cancer prevention and that this problem stems in part from the values of powerful interest groups. For example, philosopher Kristin Shrader-Frechette argues that major promotional events like Breast Cancer Awareness Month are inappropriately skewed toward promoting research for cancer cures rather than cancer prevention. She explains that the primary funder for Breast Cancer Awareness Month, AstraZeneca, is a multinational company that produces both the chemotherapy drug Tamoxifen and a number of chlorinated chemicals that appear to cause breast cancer. Therefore, she says it is no wonder that AstraZeneca pushes the slogan, "Early detection is your best prevention." Efforts to detect and treat cancer are beneficial to AstraZeneca's bottom line, whereas she contends that the company would prefer to avoid asking questions about how we could prevent cancer by limiting our exposure to industrial chemicals.

It is striking to see that very similar issues about the choice of questions arise in the research on industrial pollution that we discussed earlier in this chapter. A great deal of scientific effort is currently focused on determining the probability that people will suffer ill effects, such as getting cancer, based on their current exposures to toxic chemicals. Just as critics of contemporary research on cancer argue that we should shift our research questions to focus more on prevention, many critics of contemporary pollution research argue that we should focus more effort on finding safer alternatives to potentially hazardous chemicals.

For example, philosopher Carl Cranor has argued that our typical regulatory approaches have encouraged companies to spend decades disputing the federal government's risk assessments of their products, whereas we should instead be designing policies that incentivize them to look for safer alternatives. Cranor has emphasized that the US Toxic Substances Control Act (TSCA), which was originally passed in the 1970s, allowed companies to put chemicals on the market without showing that the chemicals were safe. Government agencies then faced a significant burden of proof to force companies to take potentially dangerous chemicals off the market. Thus, this legislation implicitly encouraged companies to ask how they could generate scientific support for keeping their products on the market rather than asking how to produce safer products. In turn, these research questions tended to support some values (e.g., the economic growth of chemical companies) rather than other values (e.g., public health). In 2016, TSCA was amended to give the Environmental Protection Agency (EPA) more power to approve chemicals before they could be marketed. Although the bill did not include as much financial support for research on safer chemicals as environmental groups would have liked, it will hopefully still provide greater incentives for companies to develop more environmentally friendly products.

A dispute over the use of chlorinated organic compounds in the Great Lakes region of the United States and Canada provides a concrete example of the general problems described in the previous paragraph. In the early 1990s, the environmental organization Greenpeace and a binational advisory committee called the International Joint Commission argued that the entire class of synthetic chlorinated organic chemical compounds should be phased out of industrial use to the greatest extent possible. They argued based on their experiences of chemical contamination in the Great Lakes that these chemicals tend to be highly toxic and persistent in the environment, and typical risk assessment approaches are not able to systematically distinguish in a timely manner between those that are toxic and those that are not. Rather than focusing scientific effort on trying to assess which chemicals in the class were toxic at particular dose levels, they argued that the values of promoting public and environmental health could be better supported by focusing researchers' attention on developing alternatives to chlorinated compounds. The chlorine industry argued that these proposals were unscientific and economically catastrophic, and the call for a phase-out did not gain political traction.

When investigating complex topics like depression or cancer or pollution, researchers face numerous choices about what questions to ask. Choosing to ask some questions rather than others is not obviously problematic in the same way as, say, suppressing evidence. Nevertheless, we have seen in the cases discussed in this section that it can have equally significant consequences for society. Therefore, it is crucial to consider what sorts of values are influencing the specific questions that scientists ask about their research topics. This can be more difficult than it initially appears because the questions that scientists ask are influenced by more than their personal values; they are also influenced by the values enshrined in law and public policy. For example, as we saw in the case of depression, policies about which sorts of treatments are patentable can determine which questions receive aggressive research funding and which ones do not.

CONCLUSION

We have seen that researchers can study a research topic in many different ways, depending on their values. As summarized in table 3.1, scientists can use different methods, they can make different assumptions, and they can ask different questions. These three avenues for value influences are closely related. For example, choosing particular assumptions or questions will often influence the methods that scientists employ. Nevertheless, it is worth keeping these three avenues distinct because they do not always coincide. It is

Table 3.1 OVERVIEW OF THREE WAYS IN WHICH A TOPIC CAN BE STUDIED
DIFFERENTLY, DEPENDING ON THE VALUES OF THOSE PURSUING OR
SUPPORTING THE RESEARCH

Ways of Studying a Topic Differently	Examples
Choice of methods or research strategies	Working to alter the genetic traits of crops versus investigating agroecological farming methods
Assumptions made	Assumptions about the behaviors of those exposed to toxic chemicals (e.g., how much fish is consumed or what parts of the fish are consumed)
Specific questions asked	Looking for drugs that alter neurochemistry versus exploring the impact of exercise on depression

also worth maintaining a distinction between the choice of research questions (discussed in this chapter) and the choice of research topics (discussed in chapter 2). While these two activities blur together in many cases, it is extremely important to recognize that even after a particular topic has been chosen for investigation, one can study it in many different ways. As we have seen in this chapter, the decision to pursue one line of investigation rather than another can significantly advance some values over others.

Consider again the case of medical research. The United States is currently struggling to deal with extremely high medical expenses that are predicted to soar even higher in the future. Future research can either exacerbate or ease these expenses, depending on whether the medical field focuses more effort on investigating relatively inexpensive, preventive approaches or expensive, patented drugs and treatments. Of course, individual researchers are limited in how much they can steer the field based on their own values. The direction of medical research is determined in large part by the decisions of government agencies like the National Institutes of Health and by the market forces that influence pharmaceutical companies. These entities, in turn, are influenced by a wide range of other factors. For example, pharmaceutical companies are heavily influenced by the patent policies and insurance reimbursement policies that determine how much they can charge for the different sorts of treatments that they investigate. Chapters 2 and 7 consider avenues for influencing these social structures that determine which methods are used, which assumptions are made, and which questions are asked.

SOURCES

More information about golden rice is available in Potrykus (2002), Pringle (2003), Shiva (2002), and Ye et al. (2000). For more information about the IAASTD report, see Elliott (2013b), IAASTD (2009), and Stokstad (2008). The strengths and weaknesses of genetically modified crops are discussed in NAS (2016). Regarding contemporary agricultural questions and agroecological research strategies, see Lacey (1999), Perfecto et al. (2009), and Shiva (1988, 1991). For the benefits of agroecological and organic strategies to women, see Lyon et al. (2010). For the tendency to focus on technical questions rather than questions from the social sciences, see Dahlberg (1979) and Patel (2007). Patel describes US policies toward India during the twentieth century, and Schlesinger and Kinzer (1999) describe the coup orchestrated by the United States in Guatemala.

For classic philosophical work on the role of background assumptions in scientific reasoning and the importance of developing appropriate venues for evaluating these assumptions, see Longino (1990, 2002). For further discussion of the assumptions and methodological choices involved in risk assessment, see Cranor (2011), Raffensperger and Tickner (1999), Shrader-Frechette (1991), and Wynne (2005). Brown and Mikkelsen (1990) discuss the Woburn case and the concept of popular epidemiology. For information about MEJO, see Powell and Powell (2011). Information on the differences between the assumptions and methodological choices made by many citizen groups and many experts can be found in Brown and Mikkelsen (1990), Raffensperger and Tickner (1999), Shrader-Frechette (1991), and Wynne (2005). For scientists' responsibilities to consider the social consequences of their work, see Douglas (2003, 2009), and Elliott (2011b).

The books mentioned in the chapter that are critical of the pharmaceutical industry include Angell (2004), Goldacre (2012), Gotzsche (2013), and Kassirer (2005). Turner et al. (2008) reports on the failure of negative studies about antidepressants to be published. Brown (2002), Goldacre (2012), and Healy and Catell (2003) tell the stories of Prozac, Paxil, and Zoloft. Overviews of the influences of pharmaceutical companies on biomedical research are provided by Holman (2015) and Sismondo (2007, 2008), and correlations between funding sources and the results of biomedical research are discussed in Bekelman et al. (2003). For further discussion of medical research concerning depression, see Musschenga et al. (2010). Shrader-Frechette (2007) discusses cancer research and the focus on cure rather than prevention. Frickel et al. (2010) provides an introduction to debates about how to handle organochlorine pollution in the Great Lakes region.

CHAPTER 4

What Are We Trying to Accomplish?

By all accounts, Dave Rosgen is a fascinating character. Raised on a ranch in Idaho, he became a forester there with the US Forest Service. He grew concerned that logging and road-building were damaging nearby rivers and streams, and he set out to study this damage in detail. As he worked, he observed that the disturbance to the streams depended significantly on their individual physical characteristics. Following up this insight, he began to collaborate with Luna Leopold, a prominent scientist at the University of California, Berkeley, who had formerly led the US Geological Survey. With Leopold's help, Rosgen developed a system for classifying rivers into a few major categories based on features like their width, depth, and slope. This work ultimately led him to the insight that made him famous in the river restoration community—he could use his classification system as a guideline for predicting how rivers and streams would respond to damage and how they could best be restored.

Unfortunately, some of the qualities that made Rosen's classification system so popular have also contributed to controversy about it. In particular, Rosgen celebrated the fact that his approach was relatively easy to learn and simple to use. One did not need an extensive geological education in order to use it as a guide to restoring rivers. The corresponding problem, however, is that it is not as reliable as the extensive analyses that geological experts could provide. Thus, it has the potential to generate misguided, poorly designed restoration efforts that ultimately fail.

The controversies over Rosgen's approach illustrate the main theme of this chapter, which is that values have an important role to play in setting the aims of scientific inquiry in particular contexts. It might initially seem like the aims of science are fairly obvious; one might think that scientists are always focused on working with the most reliable, accurate theories or models that they can develop. When one looks more carefully at scientific practice, however, one finds that scientists can have different aims in different contexts. Sometimes they might indeed be trying to be as accurate as possible, but it might be difficult to model everything in such great detail. Thus, they might have to decide which features of the world they want to model with extra care. In other cases, especially when working with regulators or policymakers, scientists might find that accuracy is only one of multiple goals; in these contexts, it might be particularly important to develop methods or models that can generate results relatively quickly and inexpensively. Similarly, when collaborating with citizen groups, scientists are sometimes asked to address very complex questions that cannot easily be answered using the most precise scientific methods available; instead, they might have to try novel, interdisciplinary, or experimental methods that are regarded by the scientific community as less than ideal. In still other cases, scientists might be focused on developing and investigating the characteristics of new models or theories without worrying too much for the time being about whether they will ultimately be good enough to adopt as an established part of science.

In this chapter, we will focus on three examples to illustrate the many different contexts in which scientists make decisions about the aims of their work. First, we will use Rosgen's research to illustrate how scientists working with regulatory agencies or policymakers can choose practical aims (such as developing models that generate results quickly and inexpensively) that go beyond merely arriving at the most accurate information. Second, we will examine theories about human evolutionary development to show that scientists can sometimes aim to develop or explore theories that they would like to make available to the scientific community for further investigation. Third, we will look at climate modeling to highlight the decisions that scientists have to make about what qualities they want their models to have. In each of these three examples where scientists are choosing the aims of inquiry—setting practical aims other than truth or reliability, deciding which theories or models are most important to explore, and deciding what qualities their models should have—values have an important role to play.

REGULATORY AGENCIES AND THEIR PRACTICAL AIMS

Rosgen referred to his new system for categorizing rivers as Natural Channel Design (NCD). It turned out to be very popular because it provided a relatively straightforward roadmap for performing restoration projects. Before long, Rosgen developed a training and certification program, including courses that cost hundreds or even thousands of dollars to attend. His approach soon became the dominant model for river restoration in the United States, and many regulatory agencies began to require those who bid on restoration projects to be certified in his approach. As recounted in a piece in the journal *Science*, the appeal of his courses stemmed in part from his unique style of communicating. Rosgen celebrated the fact that his relatively simple approach enabled users to avoid the complexity of the "high puke-factor equations" they might otherwise have to use.[1] He used phrases like, "Don't be a pin-headed snarf. . . . Read the river!"[2] And he did not come from an academic background. Although he ultimately published his classification system in a respected academic journal and eventually earned a PhD, his approach originated from a practical perspective rather than a scholarly one.

As noted in the introduction to this chapter, the practical focus of NCD has contributed to controversies over its legitimacy. Some academics have found it particularly irritating that they have been forced by regulatory agencies to obtain certification in Rosgen's system even though they have advanced training in technical fields relevant to stream restoration. These concerns cannot be attributed solely to an attitude of sour grapes. Some scholars have pointed out that his system involves oversimplifications that can result in failed restoration projects, and he has been accused of abandoning careful quantitative analyses for nonquantitative "geomagic."[3] It is difficult to determine how often his approach is actually likely to fail, however. Critics have pointed out cases where they think his simplified analysis would give incorrect results, but Rosgen typically responds that the critics obtained the incorrect results because they did not use his approach properly.

Along these lines, Rosgen argues that some of his students have engaged in unsuccessful restoration projects because they did not follow the details of his technique. As recounted in *Science*, his students, sometimes called "Rosgenauts," can easily become overconfident. Matthew Kondolf, a scientist at the University of California, Berkeley, contends that "it's deceptively accessible; people come away from a week of training thinking they know more about rivers than they really do."[4] According to restoration consultant Scott Gillilin, "students come out of Dave's classes like they've been to a tent revival, their hands on the good book, proclaiming 'I believe!'"[5] In response, Rosgen has tried to develop further courses to better train his disciples, but Gillilin still worries that "it's becoming a self-perpetuating machine; Dave is creating his own legion of pin-headed snarfs who are locked into a single approach."[6]

As noted in the introduction to this chapter, the conflicts that have emerged over Rosgen's NCD approach to stream restoration illustrate another

1. Malakoff 2004, 938.
2. Malakoff 2004, 937.
3. Malakoff 2004, 938.
4. Malakoff 2004, 939.
5. Malakoff 2004, 939.
6. Malakoff 2004, 939.

important way in which values enter science. Namely, scientists sometimes have specific practical aims that they are trying to achieve with their methods and models, especially when they are working with regulators or policymakers. For example, they may want to develop models that are not overly complicated, that do not take too long to yield results, that can be used across regulatory agencies in a standardized way, or that are particularly good at predicting specific policy-relevant pieces of information. These serve as additional aims that can extend and sometimes stand in tension with the typical scientific aim of obtaining maximally reliable information.

In the case of stream restoration, for example, government agencies want scientific methods that are "rapid, reliable, and repeatable."[7] In chapter 1, we saw that the reason people typically deny that values are relevant to science is because values seem irrelevant to determining whether a model or theory is true or predictively accurate. But in practical situations like this one, scientists are called upon to achieve more than just accuracy or reliability; they are asked to provide methods that are rapid and repeatable as well. Thus, values are relevant to deciding how to prioritize and balance these different aims. Part of why Rosgen's approach is so popular is that it fulfills the values of being easy to use, easy to teach, and easy to apply in many different circumstances. In many cases, a fluvial geomorphologist might be able to provide technically sophisticated models that would yield somewhat more accurate results than the NCD approach, but those highly accurate models would often be less successful at achieving the other aims that regulators have. Thus, scientists and regulators have to reflect on how to weigh these conflicting values.

Regulating Industrial Chemicals

Much like the case of Rosgen and NCD, the philosopher Carl Cranor has described another regulatory situation where scientists and regulators have to appeal to values in order to specify their aims. Cranor's analysis illustrates how they can reason carefully about what kinds of methods or models they should be aiming for. He points out that the procedures that regulatory agencies use for assessing risks from toxic chemicals are typically very slow and labor-intensive. For example, he notes that organizations like the International Agency for Research on Cancer (IARC) and the National Toxicology Program (NTP) have identified a number of substances that appear to be carcinogenic. Nevertheless, only a small percentage of those substances have actually been regulated because government agencies simply do not have the time or human

7. Personal communication via email from Rebecca Lave, February 4 and February 8, 2013, based on interviews with staff from the Army Corps of Engineers.

resources to perform detailed assessments of these chemicals. This is an ongoing and very serious problem because the chemical industry adds hundreds of new chemicals to the market every year, and for most of them we have very little information about their toxicity.

Cranor describes an experiment that the California Environmental Protection Agency (CEPA) performed in an effort to address this problem. During the five years prior to the experiment, the CEPA had been able to assess only 70 chemicals. Officials at the CEPA identified a much simpler methodology that they could use for assessing the likely risks of potentially toxic chemicals. It was much quicker and easier to use, but it was slightly less accurate than the traditional, more detailed approach. According to Cranor, the expedited approach significantly over-predicted the toxicity of chemicals about 3% of the time (in comparison with the traditional approach), moderately over-predicted toxicity about 12% of the time, and moderately under-predicted about 5% of the time. But this decreased accuracy does not seem like an unreasonable price to pay, given that the CEPA was able to analyze 200 chemicals in only eight months.

A critic might still question whether it really does make sense to value the rapidity of the expedited approach more than the accuracy of the traditional approach. By modeling the overall financial costs for society of using the traditional versus the expedited approach, Cranor found that it does indeed make more sense for society to place greater value on the rapidity of the expedited approach. On one hand, when toxic substances go unregulated (because regulatory agencies have not had time to assess their risks), this imposes costs on society in the form of healthcare, illness, and death. On the other hand, when relatively harmless substances are overregulated, this imposes costs on society in the form of lessened economic activity. Thus, the CEPA's expedited approach generates costs for society as a result of over-predicting the toxicity of some chemicals, but it also saves society money because it speeds up the regulatory process and allows regulators to pull toxic chemicals off the market more quickly.

In order to strengthen his analysis, Cranor pointed out that economists generally consider the financial costs of letting a toxic chemical go unregulated to be about 10 times as great as the financial costs of regulating a chemical that is harmless. Therefore, given that the CEPA's expedited approach was almost as accurate as the traditional approach and much more likely to help regulators pull toxic chemicals off the market, it is not surprising that Cranor found the expedited approach to be financially advantageous for society. What is more striking is that it still turned out to be financially preferable even if the costs of letting a harmful chemical go unregulated were considered equal to the costs of regulating a harmless chemical. Even more surprising was Cranor's finding that the expedited approach would still make more financial sense even if it were vastly less reliable and ended up regulating harmless chemicals much more often. Thus, Cranor's analysis shows that when society has economic values like saving money, it can make sense to prioritize speed and ease of use over accuracy when evaluating scientific methods and models.

Of course, this finding does not give regulators carte blanche approval to use whatever sloppy techniques they prefer. Scientists, regulators, and concerned citizens need to work together to determine which values are most important and therefore what their practical aims should be when developing methods and models for regulatory purposes. This is important because there could be some cases in which policymakers are tempted to use sloppy techniques that are not optimal for fulfilling social values. For example, regulators have also sought to employ expedited approaches for evaluating the quality of wetlands, and it is possible that they have ended up going somewhat too far in that case.

Regulating Wetlands

For most of American history, wetlands were regarded as something to be avoided and eliminated. Over the past 300 years, more than half of the wetlands present in the United States were destroyed. Given the association between wetlands and the malaria epidemics discussed in chapter 2, it is perhaps not surprising that the early American colonists had a decidedly negative attitude toward them. But wetlands were not just regarded as sources of illness; they were also regarded as evil and spiritually corrupt places. Classic literary works like the *Divine Comedy, Beowulf,* and *Pilgrim's Progress* cemented this attitude by portraying swamps, marshes, bogs, and fens as places inhabited by wickedness and sin. It did not help that Native Americans were able to use swamps to their advantage during military campaigns against European colonists, and wetlands were regarded as a nuisance by farmers who wanted to get the full productivity out of their land.

By the middle of the twentieth century, however, people were beginning to recognize that wetlands could have valuable qualities as well as drawbacks. Hunters, bird lovers, and fishermen discovered that wetlands were important for maintaining bird and seafood populations, and scientists found that they played important roles in the purification and recharge of ground water as well as in flood control. In fact, as chapter 6 discusses, the development of the term "wetland" stems from this time period, when many scientists and environmentalists were trying to promote more positive public attitudes toward these lands that had been dismissed as unimportant "swamps" in the past.

By the late 1980s, when George H. W. Bush was running for president, wetlands were regarded as sufficiently important that he campaigned on a policy of "no net loss." He regarded this as feasible because wetlands that were lost to development could be replaced by wetlands that were restored or created. This policy ultimately proved to be very difficult to enact, and even though every subsequent president has paid lip service to this policy, it does not appear that any of them actually succeeded. In fact, as chapter 6 discusses, the administration of President Bush was even accused of changing the technical regulatory definition of wetlands so that fewer lands ended up falling under the definition and needing to be preserved.

Nevertheless, the "no net loss" policy has still been an important principle, and one of the most important regulatory tools for pursuing this policy has been section 404 of the Clean Water Act. It requires the Army Corps of Engineers, in consultation with the Environmental Protection Agency, to provide permits to those who want to dredge or fill wetlands on their property. It is very uncommon for these agencies to deny permits outright, but they often require that mitigation measures be taken. This mitigation frequently takes the form of preserving or restoring wetlands in other locations to make up for the wetlands that are destroyed. As a result of this regulatory policy, a market has developed for "wetland banking."

Wetland banking occurs because regulators are not particularly inclined to force developers to preserve or restore wetlands themselves. Instead, it is more appealing to require that they purchase mitigation "credits" from other companies that specialize in creating "banks" of preserved or restored wetlands. Billions of dollars are now spent each year on wetland mitigation, and there are hundreds of banking sites. However, the scientific challenge for regulators is that they have to compare the wetlands that are destroyed and the wetlands that are preserved or restored to make sure that they are sufficiently similar to justify trading the two.

Geographer Morgan Robertson has analyzed the process of wetland banking and has emphasized that the scientific techniques used in this context are very different from those that would be best if one were aiming to provide the most ecologically accurate comparisons. Much like in the case of stream restoration, this happens because regulators are trying to achieve comparisons that are not only accurate but also relatively cheap and quick. Robertson reports that scientists have worked with regulators to develop "rapid assessment methods" (RAMs) that they can use to evaluate the quality of wetlands. These RAMs yield numerical scores that are supposed to provide rough indications of wetlands' abilities to fulfill functions like providing flood control or bird habitat. Nevertheless, they tend to be very crude. In addition to reducing the complex features of a wetland to a single number, they draw on very limited information. For example, they are often based on data about the plants that are present because plants are relatively easy to observe. But even these limited data are less accurate than they could be. Robertson points out, for example, that according to regulatory guidelines, plant species are often supposed to be identified in May or June. Unfortunately, this is a time of the year when many plants have not yet flowered and are therefore difficult to categorize.

These crude techniques have come under fire from some ecologists. They argue that these approaches for comparing wetlands can lead to serious losses in wetland functions. When politicians and regulators hail their successes at replacing wetlands destroyed by development with newly restored wetlands, their enthusiasm may be deeply misguided. If the new wetlands do not provide the same important functions that the destroyed wetlands provided, there is little reason to be enthusiastic about replacing one with the other. Thus, in this case the inaccurate assessment methods employed by regulators may be contributing to socially undesirable outcomes. To make the argument of the ecologists totally compelling, however, we would need to examine the social consequences of using more ecologically accurate methods. If those methods took so long to employ that it was no longer possible to engage in wetland banking, and if this prevented wetlands from being restored or preserved, then it might be better to employ the crude techniques currently in use. But if we could employ more accurate ways of comparing wetlands and still maintain wetland banking or some other system for preserving wetlands, perhaps this would in fact be best for fulfilling our social values.

From the cases discussed in this section, we can see that it sometimes seems appropriate to prioritize aims like speed and ease of use when assessing scientific methods and models, whereas sometimes this seems to be problematic. How can we tell the difference? The key—as we have seen throughout this book—is to clarify the overall goal that we are aiming to achieve. Cranor argued in the case of chemical risk assessment that both economic and public health values are best served by employing quicker, less accurate methods. In contrast, we saw that some ecologists think our techniques for comparing wetlands are so inaccurate that they are not serving our environmental values that call for preserving ecosystem services. We can see in both cases that values are relevant for helping scientists decide which aims to prioritize when choosing methods and models.

CHOOSING THEORIES TO PURSUE

A second example of a context in which scientists need to choose their aims involves decisions about which theories or models are most important to *develop* or *pursue* or *explore*. When most people think about scientific reasoning, they typically envision scientists deciding what theories they should *believe* or *accept* as true. A central aspect of scientific practice, however, is deciding which preliminary ideas are most worthy of being pursued further, even if it is not yet clear how accurate or reliable they will ultimately turn out to be. Values sometimes have an important role to play in setting these aims.

Consider an example from anthropology. In January 1981, C. Owen Lovejoy published a famous paper, "The Origin of Man," in the journal *Science*.

The paper presented his "male-provisioning hypothesis," which was intended to explain how upright walking, or bipedalism, evolved in humans. Over the past 100 years, anthropologists have developed a host of different explanations for the development of human bipedalism, which is highly unique among mammals. Consider some of the hypotheses that have been proposed. One hypothesis is that it assisted early humans in observing their surroundings. Another hypothesis is that it made it easier for them to use their hands. Other hypotheses abound. It might have facilitated throwing weapons, or it might have made it easier to carry infants, or it might have been related to reaching up to obtain food, or it might have been part of "threat displays" to settle disputes, or it might have helped reduce humans' exposure to solar radiation, or it might have developed as a result of wading behaviors in shallow waters.

Lovejoy suggested that the rise of bipedalism was linked to the development of monogamy. He argued that engaging in pair bonding was selectively advantageous for humans. This led to reduced conflict between males while enabling them to devote more effort to supporting their own children and less effort to other men's offspring. Coupled with this shift to monogamy, he suggested that if males left their mates to gather food and then returned—carrying the food in their arms while walking upright—it would increase their reproductive success. The activity of bringing back food would increase the commitment of women to their male providers. Moreover, this male-provisioning activity would allow women to have their babies spaced more closely together because they would have access to more food and would not have to move around as much.

Despite the appeal of Lovejoy's theory, it has been subjected to a great deal of criticism. In an extremely entertaining review of the work done by several of her fellow anthropologists, Adrienne Zihlman argued that Lovejoy's theory was just one more male-centered contribution to a long line of biased anthropological theories. She notes that in the 1960s, a variety of anthropologists developed theories focused on the theme that our male ancestors drove human evolution forward because of their hunting activities. For example, the ability to walk on two legs allegedly evolved because it enabled males to free their hands for making and using hunting tools and carrying meat. Better communication and greater intelligence evolved because of the way these abilities enabled males to plan hunts more successfully. Similarly, as Lovejoy suggested, the sharing of meat from males to females allegedly contributed to pair bonding.

Zihlman laments: "Women in prehistory were missing from these stories. The picture of ancient women was of faceless and formless beings, doing nothing, going nowhere."[8] She notes, however, that female anthropologists in the

8. Zihlman 1985, 367.

1970s began to challenge this male-centered picture of human evolution. A classic paper by Sally Slocum proposed that one could augment or replace the "Man the Hunter" account of evolution with a "Woman the Gatherer" account. For example, the same characteristics that were supposed to have been driven by the need for successful hunting could also have been driven by the need for successful gathering. These characteristics include the ability to walk on two feet, the ability to carry food, the use and construction of tools, and the intelligence to communicate and analyze the environment. Along with Nancy Tanner, Adrienne Zihlman herself then proposed an account of human evolution where women's activities played a central role and where hunting occurred much later in the evolutionary process. They suggested that molecular evidence, archeological data, and concepts from sociobiology all supported their account. According to Zihlman, the account of human evolution provided by her collaborator Tanner leaves readers with the impression of "all-powerful females, leading the way in evolution with the males tagging behind trying to keep up."[9]

Given this background, one can see why Zihlman was dismayed by Lovejoy's male-provisioning hypothesis. As she saw it, he was taking all the new evidence that gathering activities played a crucial role in human evolution, but instead of giving women the credit he flipped it around and suggested that men were doing the gathering. As she put it: "At one stroke, Lovejoy has rehabilitated Man the Hunter but by co-opting Woman the Gatherer: now it's Man the Gatherer, and Woman the Gene Receptacle is relegated to utter passivity."[10] Zihlman pulled her reference to "Woman the Gene Receptacle" from one of Lovejoy's colleagues, who argued that it was evolutionarily advantageous for each male to have its own "private gene receptacle" to mate with.[11]

From Zihlman's perspective, Lovejoy and his followers ignored all the progress that feminist anthropologists had recently made in theorizing the roles of women in human evolution; instead, he returned to a male-dominated picture. Moreover, his vision of early human behavior had a suspicious resemblance to the family structure idolized in mid-twentieth century America, with a male provider and a stay-at-home wife. As a matter of fact, Zihlman argued that, despite the "glad baritone cries" of the male science writers who were so fascinated by Lovejoy's account, it actually fit very poorly with the available evidence:

> Lovejoy's picture of the hominid female—to whom everything comes and who goes nowhere—is completely inconsistent with the mobility and activity of all known primate females, even when pregnant or carrying their offspring. . . .

9. Zihlman 1985, 371.
10. Zihlman 1985, 374.
11. Zihlman 1985, 374.

Even among gatherer-hunters like the !Kung San (Bushmen) of Botswana, the females walk for miles a day while pregnant and carrying a young infant, and manage to gather plant foods for themselves and their mates as well.[12]

This is not to say that Zihlman and Tanner got it right either. In fact, the available data are so sparse and so subject to different interpretations that it is exceptionally difficult to arrive at definitive conclusions regarding human evolution. Zihlman herself acknowledges: "Certainly women, like men, are drawing on events from their own lives and the political climate of the times. . . . Everyone engaged in building evolutionary models brings her or his own bias to the activity."[13] The question, then, is how best to respond to these inevitable biases and values. Should we just try to suppress them as best we can, or are there more fruitful ways of responding to them?

Allowing Values to Influence the Pursuit of Theories

Philosopher Philip Kitcher has suggested that, under the right conditions, it can actually be beneficial for the scientific community when individual scientists are motivated to pursue theories based on their personal values. He invites people to think back to the late 1700s, when chemists were divided between the old phlogiston theory championed by Joseph Priestley and the new oxygen theory promoted by Antoine-Laurent Lavoisier. Kitcher argues that the scientific community had a better chance of efficiently arriving at the best theory because some scientists were quickly willing to "jump ship" to Lavoisier, while others were much more stubborn and persisted in defending the old approach. Those who quickly jumped ship began developing and improving the new theory in order to determine whether it could ultimately improve upon the previous theory. But it would have been highly problematic if the entire scientific community immediately turned their attention to the oxygen theory, given that it might ultimately turn out to be a failure. Thus, on Kitcher's view, it was very valuable for scientists to take differing perspectives on which theories were most worthy of pursuit.

With this in mind, Kitcher argues that even when the scientific *community* is focused primarily on pursuing the truth, this goal can be well served when *individual* scientists are motivated by considerations other than truth. These motivations could include economic or ethical or political values or even advancement in one's career. For example, an individual scientist might recognize that a highly speculative new theory is fairly unlikely to succeed.

12. Zihlman 1985, 374.
13. Zihlman 1985, 375.

Nevertheless, she might also recognize that if she were to be one of the few scientists who developed the theory further and it did ultimately succeed, she would receive a great deal of fame and notoriety. Other scientists who are more risk-averse might conclude that they can develop a more personally appealing career by working on a theory that has more evidence in its favor. According to Kitcher, the scientific community's goals of pursuing the truth might be very well served by having these scientists pursue different theories based on a range of values. By dividing up the "cognitive labor" among a variety of different scientists, the scientific community can pursue a number of potentially promising theories while also maintaining a good deal of attention on the best currently available theory.

Unfortunately, this approach can run into problems when the entire scientific community has particular biases or values that are not adequately counterbalanced. For example, consider Zihlman's worries about Man the Hunter theories of human evolution. As Zihlman pointed out, the entire anthropological community in the 1960s was failing to ask important questions about the roles that women might have played in the evolutionary process. Philosopher Kathleen Okruhlik points out that this sort of scenario can result in very problematic influences of values on science, even if individual scientists are not consciously influenced by values. For example, if all the available anthropological theories that have been proposed are focused on male contributions to human evolution, it is likely that the anthropological community will end up adopting a male-centered theory, even if the male-centered theory is incomplete or imperfect. This is a very important way in which undesirable values can influence science, even if nobody is purposely trying to steer science in this manner. Thus, it is important to find ways to evaluate and perhaps counteract these sorts of subconscious value influences.

Bringing Kitcher's and Okruhlik's insights together, we can identify another important role for values in science. Namely, scientists may decide to pursue particular models or theories because they promote values that the scientists or other stakeholders think should be better represented in science. So, for example, even if a "woman the gatherer" theory did not seem particularly likely to be true, it might be very reasonable for a feminist anthropologist to pursue the theory anyway. If most of the competing theories focused on male contributions to evolution, and if it seemed that those theories tended to have negative effects on society's views about women, it might be reasonable to explore whether any woman-centered theories could be developed. Or, she might engage in a research project focused on identifying as many weaknesses of the male-centered theory as possible so that it would not gain widespread acceptance without adequate scrutiny. This is not to say that the male-centered theory would ultimately be rejected by the anthropological community or that a feminist theory would ultimately obtain enough evidence in its favor to be accepted, but it might be good use of a feminist's time to make sure that woman-friendly theories at least received an adequate hearing.

Some Objections

One might worry, however, that this proposal sounds much too similar to the situation surrounding Lysenko and Vavilov that we discussed in chapter 1. Are we not coming perilously close to the problem of wishful thinking, which occurs when scientists choose which theories to accept and reject based on what they want to be true? To respond, we need to return to an idea that we have used throughout this book: values have an appropriate role to play in science if they help scientists to achieve their legitimate goals. If a group of scientists is trying to decide whether a theory is likely to be *true* or *reliable*, then values are typically not relevant to answering this question. However, if scientists are trying to decide which theory is worth *developing* or *investigating* or *critiquing* further, then values might indeed be relevant. As Kitcher indicated, it can be entirely reasonable for a scientist to investigate a theory not because it seems most likely to be true but rather because the scientist would like to make it available to the scientific community as an eventual competitor to the theories that are currently dominant.

If scientists do consider values relevant to their choices, however, it is crucial for them to be transparent about these influences so that they do not confuse their fellow scientists or members of the public. For example, they need to clarify whether they think a particular theory is worthy of being *believed*, or *accepted*, or *pursued*, or *hypothesized*, or *entertained*, or *developed further*, or *used for the purposes of making regulations*, or *doubted*, or flat-out *rejected*. These varying perspectives that can be taken toward a theory are sometimes called "cognitive attitudes." Different sorts of values are relevant when scientists adopt different cognitive attitudes. Values may not be relevant when scientists are deciding what to *accept*, but they can be highly relevant to deciding which theories should be *pursued*. Therefore, to prevent confusion, scientists need to find good ways to clarify their cognitive attitudes toward the theories with which they work. Otherwise, other scientists and members of the public could get the false impression that a particular theory is likely to be *true* merely because a scientist thinks the theory is *worthy of further investigation*.

One might still wonder whether it is always legitimate for scientists to pursue theories that support their preferred values. The pursuit of feminist anthropological theories is relatively easy to defend, partly because it seems to correct an imbalance in previous theories and partly because most of us would like to assist traditionally disadvantaged groups. But what if the values were less appealing? For example, consider Willie Soon, a scientist at the Harvard-Smithsonian Center for Astrophysics who has been a prominent climate change skeptic for many years. He has spent much of his career exploring the theory that variations in solar radiation might be responsible for climate change instead of greenhouse gases, and his views have been widely promoted in the climate denial movement. While it is somewhat unclear why he is pursuing a theory that has been almost universally discredited by other climate scientists, one can imagine that he might be motivated at least partly by the desire for personal advancement and perhaps also by the antiregulatory ideology that has motivated many climate skeptics (as will be discussed in chapter 5).

In a case like this one, it might not seem very justifiable for Soon's values to influence his science. But perhaps the problems in his case are not primarily related to allowing values to influence his pursuits. The most obvious problem is that, in contrast to the case of the feminist anthropologists, his solar radiation theory has already been closely examined by the climate science community and has been discredited. So he is not merely exploring a new theory that might someday prove to be a success; he is hanging onto a failed theory. Yet another worry is that he has been able to pursue his research because of $1.2 million in donations from powerful corporations and foundations that aim to sow doubt about climate change. Thus, one might worry not so much about the fact that Soon is influenced by values but by the fact that his values are being supported by an overly influential movement that represents only a narrow range of society's interests and priorities. Finally, in 2015 it came to light that Soon had not disclosed all his funding sources, even in journal articles where these disclosures were required. Thus, as we have seen throughout this book, problematic cases like Soon's can typically be attributed to the failure to meet additional conditions (e.g., transparency, representativeness, and engagement) rather than the fact that scientists allowed values to influence the theories that they pursued.

CHOOSING WHAT TO MODEL

A third example of a context in which scientists have to make decisions about their aims involves choices about what qualities their models should have. Scientific models are typically designed to represent a particular phenomenon in some way, but they can have different strengths and weaknesses. For example, a simple model might represent the major components of a phenomenon and the relationships between them but not provide enough details to make quantitative predictions about how the phenomenon will change over time. In some contexts, a simple model like this might be adequate, but in other contexts it might be important to develop a more detailed model that could facilitate future predictions. Models are always imperfect, however, so scientists still have to decide which features of the world they are most concerned to predict with their models. In some cases, they even face trade-offs, such that optimizing a model to describe some facets of a phenomenon might make it more difficult to describe other parts of it. Values often have a role to play in setting these aims, as is especially obvious in the case of contemporary climate modeling.

In 2006, for example, the British government released a massive 700-page report on the economics of climate change. It came to be called the "Stern Review" in honor of its lead author, the influential economist Sir Nicholas Stern. The report concluded that the costs of failing to take action against climate change could potentially run between 5% and 20% of the world's gross domestic product, rivaling the impact of the Great Depression and the World Wars of the twentieth century. Thus, the report argued that it would be well worth spending significant amounts of money now to cut greenhouse gas emissions in order to lessen those future costs. Political leaders like Tony Blair, the prime minister of the United Kingdom at the time, cited the report as justification for taking aggressive steps in response to climate change. Nevertheless, the Stern Review encountered a great deal of criticism.

One of the most important critics, Yale University economist William Nordhaus, challenged the report for arriving at different results than most previous economic analyses of climate change. Most previous analyses found that it made more economic sense to impose less aggressive cuts on greenhouse gas emissions now and to ramp up those cuts in the future. What is particularly striking is that, according to Nordhaus and other economists, one of the primary reasons why the Stern Review arrived at different results from most other analyses is that it employed a much lower "discount rate." Economists employ a discount rate when they compare costs and benefits at one point in time with costs and benefits at other times. For example, economists would not treat a $100 expense incurred at present in the same way as a $100 expense incurred 10 years from now. Instead, they would "discount" the value of the future expense by a particular percentage for each year that goes by before the expense is incurred. The result of this procedure is that future costs or benefits are treated as much less significant in comparison with present costs or benefits.

Choosing an appropriate percentage for discounting is a very complicated matter that can be justified based on a variety of different considerations. The most obvious reason to discount costs and benefits in the future is that if we are able to invest a dollar now it will typically be worth much more later on; therefore, having a dollar now is much more valuable than having a dollar in, say, 20 years. But there are a number of other reasons for discounting: (1) people have a psychological preference for having benefits now rather than in the future; (2) there is uncertainty about what will happen in the future, or whether the human race will even continue to exist; (3) if people in the present are less well-off or have fewer technological resources than people in the future, then money is more valuable to us now.

While the concept of discounting might appear to be fairly boring and technical, it can have shockingly large effects on policy documents like the Stern Report. When most of us think about uncertainty or disagreements related to climate models, we typically envision differences of opinion about how to model physical features of the climate. For example, we picture scientists disagreeing about how quickly the West Antarctic ice sheet will melt in response to warmer temperatures and how much the loss of ice will alter the earth's absorption of radiation from the sun. However, disagreements about discounting can have huge effects of their own. As Harvard economist Martin Weitzman put it in a review of the Stern Report, "The disagreement over what interest rate to use for discounting is equivalent here in its impact to a disagreement about the estimated damage costs of global warming a hundred years hence of *two orders of magnitude*."[14] In other words, those who employ high discount rates (thereby decreasing the appearance of future costs relative to present costs) can end up estimating the costs of climate change to be 100 times less than those who employ low discount rates! Thus, Weitzman argues that "it is not an exaggeration to say that the biggest uncertainty of all in the economics of climate change is the uncertainty about which interest rate to use for discounting."[15]

This case provides a valuable example of the very significant roles that values can play when scientists are choosing and evaluating models. Whereas most economists use discount rates of about 5% in their analyses of the economic impacts of climate change, Stern called for a discount rate closer to 1%. Therefore, because Stern was not discounting future costs very significantly in comparison with current costs, it made those costs look very significant. Most previous analyses, which effectively shrank the appearance of future costs by discounting them more heavily in comparison with present costs, concluded that it was not as important to spend significant amounts of money in the present.

Despite the significance of choosing a discount rate, there is not a straightforward fact of the matter about which rate should be employed; the choice depends on the sorts of information that people value. There are many different "truths" that scientists could calculate when they are comparing the costs of climate mitigation schemes. They could calculate how the costs and benefits of these schemes compare when future costs are discounted at 1% per year, or when they are discounted at 2% per year, or when they are discounted at 3%, and so on. Which of these truths is most important to calculate depends on

14. Weitzman 2007, 708, italics in original.
15. Weitzman 2007, 705.

our ethical values about how we ought to compare the interests of future generations relative to our own. For example, even if they were not consciously motivated by values, the economists who employed high discount rates were supporting the value of current industrial activities over values associated with future environmental health, whereas Stern's approach favored the interests of future generations.

Thus, the debate over discounting illustrates that scientists need to consider their aims (which in turn depend on their values) when deciding what they want to model and how they want to model it. When scientists create and evaluate models or theories, they have to determine what is most important to represent. As philosopher Elizabeth Anderson puts it: "Theories don't just state facts; they organize them into patterns that purport to be representative of the phenomenon being theorized, patterns that are adequate to answer some question or satisfy some explanatory demand."[16] Therefore, to determine which models or theories count as good ones, scientists have to decide what they are aiming to achieve with those models or theories. This is very similar to one of the points made in chapter 3, namely, that it is often possible to study a complex phenomenon in many different ways, so values can determine which questions one decides to ask. But this chapter places more emphasis on deciding how to *choose between* multiple models or theories of a phenomenon rather than deciding *what to ask* about it.

To provide further illustration of this point, consider some more examples from climate modeling. Philosopher Kristen Intemann examined a number of decisions that scientists need to make when developing, choosing, and using climate models, and she emphasizes that values are often relevant to making these decisions. For example, General Circulation Models (GCMs) are very helpful for predicting global average temperatures, but they are much less helpful for predicting local changes in precipitation. Regional Climate Models (RCMs) are much more helpful for this purpose. This might seem like a fairly trivial point—the choice to use a GCM or an RCM obviously depends on the questions that we want to answer. But the role of values becomes much more interesting when we consider how they can influence the development of new models.

Intemann points out, for example, that climate models incorporate a variety of assumptions and parameter values that are not completely settled by evidence. Therefore, the models need to be "tuned" by adjusting these parameters and assumptions in order to best fit the available data. However, tuning the model to better predict some phenomena (such as the distribution of precipitation across a designated region) might make the model less adept at predicting other phenomena (such as extreme weather events). Values are

16. Anderson 1995, 39.

clearly relevant to choosing which truths are of highest priority for a model to represent accurately. Similarly, our current models are better at predicting gradual climatic changes than predicting the worst case scenarios that could arise if the earth began to warm rapidly. Depending on our values, we might conclude that it is necessary to sacrifice some of the current strengths of our models so that they can better predict "worst case scenarios."

Ethical values can also influence the design of Integrated Assessment Models (IAMs). These models integrate predictions from the physical sciences with economic analyses in order to determine whether it is more economically efficient to spend money now to lessen climate change or to spend money later to adapt to climate change. Some scholars have pointed out that there are ethical reasons to structure these models so that they provide not only aggregate information about the overall future economic costs and benefits that people will experience but also distributive information about how particular groups will fare. For example, if we are particularly concerned about marginalized groups such as poor populations in low-income countries, we might think it is very important to predict how those particular groups would be affected by decisions to spend more money now rather than in the future.

Philosopher Nancy Tuana and a group of scholars working at Penn State University have argued that these sorts of decisions are so important in scientific modeling that they have coined the terms "intrinsic" or "embedded" ethics as a way of describing the issues that arise in making these decisions. Courses in research ethics typically focus on what Tuana calls "procedural ethics," namely, how to protect and share data, how to decide who should be listed as an author on a paper, how to treat experimental animals appropriately, and how to mentor students successfully. While these issues are important, Tuana suggests that we need to do a better job of training students to appreciate the nitty-gritty choices that arise in modeling and the fact that ethical values are often relevant to making these decisions. As we have already seen from Intemann's examples, these ethical values are relevant to modeling because scientists cannot represent every aspect of the world with equal accuracy when they develop a model. They have to decide what aims they are trying to accomplish in order to determine what their models should represent best. This is not to say that scientists are always cognizant of the ways their models are serving some aims better than others; as Tuana points out, they are often oblivious to the significant choices they are making. But the important point is that values are indeed relevant to these decisions, insofar as scientists are developing models that are intended to be useful for the real-world needs and concerns of decision-makers.

It is worth emphasizing that these issues of intrinsic ethics are not unique to major social issues like climate change; they are ubiquitous in scientific modeling. Philosophers Sven Diekmann and Martin Peterson recently highlighted a variety of other examples that illustrate the same sorts of choices that arise in climate modeling. For example, they noted that when the United States was preparing to perform its census in 1990, the leaders of the census had to choose between two different statistical models for analyzing the data. Both models were designed to help address the undercounting of ethnic minorities, but one was focused more on improving the data as a whole while the other was more focused on regional data. Thus, values were relevant to deciding which sorts of data were most important to improve. Similarly, Diekmann and Peterson showed that in Geographical Information System (GIS) models, various assumptions are made in order to decrease errors and increase overall predictive accuracy. However, while these assumptions improve the outputs of GIS systems in most cases, they actually hamper the successful representation of wetlands. Again, this means that values are relevant to evaluating these GIS systems because the modelers have to decide which features of the world are most important to represent accurately.

CONCLUSION

This chapter highlighted the role of values in setting the aims of scientific inquiry. The final section of the chapter used the example of climate modeling—and especially disputes over the Stern Report—to illustrate the potential for disagreements about what qualities a good model should have. After his report was released, Sir Nicholas Stern came to the United States to testify before Congress in February 2007. After testifying, he traveled to Yale University to attend a symposium focused on evaluating the report. It was particularly fitting to hold the symposium at Yale, given that some of Stern's most important critics, including William Nordhaus, were faculty members there. David Leonhardt described the setting of the symposium beautifully in an article for the New York Times: "With a couple of hundred students and professors in the audience—and a sculpture of an angry Zeus-like figure looming above the stage—the two sides went at it in the dignified, vicious way that academics do." While praising Stern as a distinguished scholar, Nordhaus allegedly accused him of committing "cruel and unusual punishment on the English language," and another Yale economist, Robert O. Mendelsohn, compared Stern to the Wizard of Oz. For his part, Stern insisted that "I've still not heard a decent ethical argument" for adopting the higher discount rates preferred by Nordhaus and Mendelsohn.

These academics may have been a bit more dramatic than necessary with their criticisms, but the effort to subject the Stern Report to a careful analysis of its underlying values is exactly the sort of activity that this book aims to promote. As Leonhardt confirmed in his article, Stern's critics had some concerns about other aspects of the report, but their fundamental objection was to his choice of a low discount rate. This dispute is particularly beneficial insofar as the underlying value-based disagreements were brought to the fore so they could be discussed in a public setting. In many cases, modelers make important choices that are never thoroughly discussed, and they do not even realize that their models serve some aims (and their accompanying values) better than others.

The first two sections of the chapter looked at two other examples of contexts in which scientists need to think about their aims. First, when they work with other stakeholders, such as regulators or policymakers, they may need to adopt practical aims (such as minimizing expenses or generating results quickly) that those collaborators bring to the table. Second, in some contexts scientists need to decide what sorts of models or theories they would like to develop and make available to the scientific community for further investigation. In these cases, scientists are sometimes not merely oblivious to the role of values; they may even attempt to suppress or eliminate them entirely.

Recall that Carl Cranor argued for employing somewhat less accurate risk-assessment methods because they generate results much more quickly and inexpensively than more accurate ones. Even though it makes good sense to employ the methods that best serve our social aims of minimizing economic costs and protecting public health, scholars who argue for less accurate methods can easily come under attack for promoting "bad science." Sometimes citizen groups are also criticized when they call for action to address public-health threats even when they have not been able to employ the most high-quality methods available to determine the extent of the threats. It is easy to forget that science, like any activity, cannot be labeled "bad" or "good" without considering what we are trying to accomplish with it. As we have seen throughout this chapter, accuracy is one of the most important aims of science, but it is often not the only one. Thus, scientists need to decide what aims are most appropriate in particular research contexts, and values are often relevant to making those decisions. The next chapter turns to a closely related issue, namely, the question of how to decide when enough evidence has been gathered in order to draw conclusions on the basis of limited information.

SOURCES

Malakoff (2004) provides a fascinating overview of Dave Rosgen's life and work. Lave (2009) provides a very thoughtful analysis of the conflict between Rosgen and his critics. Cranor (1995) provides his analysis of the CEPA's expedited risk-assessment procedures, and Elliott and McKaughan (2014) provide an overview of both the CEPA case and the wetland banking case. Further information about wetlands and wetland banking can be found in Elliott (2017), Hough and Robertson (2009), and Robertson (2006). For more discussion of the idea that the goals of science in a particular context determine what methods or models are appropriate, see Elliott (2013a), Elliott and McKaughan (2014), and Intemann (2015). Brown (2013) and Potochnik (2015) also discuss the diverse aims of science and the ways that values become relevant as a result. O'Fallon and Finn (2015) emphasize the differences between the research aims that often arise within community-originated research efforts versus the research aims typically associated with academic research.

Zihlman (1985) provides a fascinating review of anthropological research on human evolution. Niemitz (2010) summarizes numerous hypotheses concerning the development of bipedalism, and Lovejoy (1981) introduces the male provisioning hypothesis. For additional perspective on feminist research in anthropology and archeology, as well as the concepts of "man-the-hunter" and "woman-the-gatherer," see Longino (1990) and Wylie (1996). Elliott and Willmes (2013) discuss Lovejoy's male-provisioning theory as well as the different cognitive attitudes that can be taken toward scientific theories (see also Elliott 2013a, Elliott and McKaughan 2014, and McKaughan and Elliott 2015). Kitcher (1990) develops the idea that the scientific community can be well served even when individual scientists are not focused solely on uncovering the truth. The notion that scientists will choose the best among the available theories, even if they are all influenced by values, is in Okruhlik (1994). Willie Soon's research on climate change is discussed in Gillis and Schwartz (2015).

For information about discounting and the debate surrounding the Stern Report, see Leonhardt (2007), MacLean (2009), Nordhaus (2007), Stern (2007), and Weitzman (2007). Elizabeth Anderson's views about the role of values in evaluating scientific theories and models can be found in Anderson (1995). Intemann's description of values in climate modeling can be found in Intemann (2015), and an example of Tuana's call for intrinsic ethics is in Schienke (2011). Diekmann and Peterson (2013) and Potochnik (2015) provide further examples of the role of values in modeling. Wendy Parker has published a great deal of important work about climate modeling, and her views about determining whether models are adequate for particular purposes can be found in Parker (2009).

CHAPTER 5
What If We Are Uncertain?

During congressional hearings in June 1988, climate scientist James Hansen took a stronger stand than many of his fellow scientists and confidently declared that, based on his analysis of the available evidence, the earth was warming as a result of the greenhouse effect. He affirmed: "Global warming has reached a level such that we can ascribe with a high degree of confidence a cause and effect relationship between the greenhouse effect and observed warming.... It is already happening now."[1] For decades, scientists had been investigating the possibility that the increased atmospheric concentrations of greenhouse gases caused by burning fossil fuels might alter the global climate. Basic physical principles suggested that the changes that humans were causing to the atmosphere would probably end up warming the earth. In the late 1980s, however, many climate scientists did not feel that there was enough evidence to assert with confidence that increasing atmospheric concentrations of greenhouse gases were already causing global warming.

This chapter examines the difficult issues that scientists like Hansen face when they are dealing with uncertainty. The first section uses Hansen's situation to show that values are relevant to scientists when they have to decide how boldly to communicate about uncertain scientific findings to the public. This involves difficult decisions about whether or not to draw straightforward conclusions, how much to hedge those conclusions with qualifications and uncertainties, and how to frame those conclusions with respect to broader social issues. The second section of the chapter focuses closer attention on one of these issues; it shows that values are relevant for deciding how much evidence scientists should demand in order to draw conclusions.

1. Shabecoff 1988.

The first two sections of the chapter take uncertainty as a given and argue that values are relevant for responding to it. The third section flips this relationship around and argues that values can actually generate or contribute to uncertainty. When scientific research conflicts with people's deeply held values, they are often motivated to challenge the science in every way possible. This has been happening in a number of recent cases, including in research related to climate change, genetically modified foods, evolutionary theory, and vaccines. The third section of the chapter explores how values can contribute to uncertainty and considers how to decide when this is appropriate and when it is not.

COMMUNICATING ABOUT UNCERTAIN FINDINGS

When Hansen testified to Congress, he was the Director of NASA's Goddard Institute of Space Studies. He was also leading a team of modelers that was becoming increasingly confident that climate change was happening and that it could generate extreme weather events such as heat waves and droughts. Hansen decided that he had a social responsibility to speak out confidently about the team's findings. According to his own account of the situation, he "weighed the costs of being wrong versus the costs of not talking,"[2] and he concluded that it was best to draw greater public attention to the existence of climate change and its possible effects. This is precisely the sort of value-laden decision that we want to explore in this section. When scientists are working on important public issues and have incomplete scientific information, they have to decide how boldly to present their findings. On one hand, scientists can remain very cautious and present their data in modest ways that minimize their chances of saying anything misleading. On the other hand, they can act more like public advisers or even advocates and highlight the potential ramifications of their findings for the public. There are advantages and disadvantages to each approach, as well as to the range of options in between these approaches. Deciding which approach is best in a particular context requires weighing these advantages and disadvantages, and this requires value judgments.

In Hansen's case, he had already given congressional testimony in 1986 and 1987, but the popular press paid him very little attention. Hansen's testimony in 1988 occurred during the summer, in part because congressional staffers thought that it would be more compelling to introduce legislation about climate change during that time of the year. In his testimony, he presented his conviction that the earth was getting warmer, that we could

2. Weart 2014.

probably associate that temperature increase with the greenhouse effect, and that his models predicted increased frequencies of droughts as a result. Given that the United States was facing both record temperatures and a drought that summer, Hansen's testimony got extensive attention. His statements were featured in front-page newspaper stories as well as television and radio shows.

In addition to the media attention that he received, he also faced criticism from a number of his fellow scientists. In 1989, Richard Kerr wrote a News piece for the journal *Science* that reported on Hansen's testimony and the subsequent responses to it. According to Kerr, he "irked practically everyone in the field."[3] Michael Schlesinger, a modeler at Oregon State University, complained that "his statements have given people the feeling the greenhouse effect has been detected with certitude. Our current understanding does not support that. Confidence in detection [of the greenhouse effect] is now down near zero."[4] Tim Barnett, a scientist at Scripps Institution of Oceanography, insisted that "the variability of climate from decade to decade is monstrous. To say that we've seen the greenhouse signal is ridiculous. It's going to be a difficult problem."[5] Alan Robock, from the University of Maryland, explained that "what bothers a lot of us is that we have a scientist telling Congress things we are reluctant to say ourselves."[6] Hansen's colleagues worried that by making overly confident claims in such public venues, he risked sabotaging people's trust in climate scientists. As Danny Harvey of the University of Toronto summed it up, "Jim Hansen has crawled out on a limb. A continuing warming over the next 10 years might not occur. If the warming didn't happen, policy decisions could be derailed."[7]

It might be tempting to characterize this situation as one in which Hansen allowed values to influence his work in a questionable way, whereas his critics rightly refrained from incorporating values in their work. Even if one disagrees with Hansen's actions, however, it would be misleading to claim that he was the only scientist who incorporated values in his decision-making. The quotation from Danny Harvey illustrates that other climate scientists were also seeking to communicate information in socially beneficial ways. Harvey challenged Hansen by insisting that Hansen's confident approach to communicating information could diminish public trust in climate scientists, thereby leading to poor policy decisions in the future. Thus, Harvey and other climate scientists were not excluding values from their work; rather, they were prioritizing different values—they thought it was much more important to maintain trust in the scientific community than to highlight extreme events

Values present from both parties, but priorities were different.

3. Kerr 1989, 1041.
4. Kerr 1989, 1041.
5. Kerr 1989, 1043.
6. Kerr 1989, 1043.
7. Kerr 1989, 1043.

that might not come to fruition. As Richard Kerr reported in 1989, most climatologists at that time agreed with Hansen that greenhouse gas emissions were probably contributing to global warming. They just thought that it was socially irresponsible, given the uncertainties that they faced, to make bold statements that could make it more difficult for scientists to influence Congress in the future. In contrast, Hansen weighed the pros and cons differently and concluded that the most socially responsible approach was to alert the public to extreme weather events that they could face. He told reporters after his testimony that it was important to "stop waffling, and say that the evidence is pretty strong that the greenhouse effect is here."[8]

Reflecting on Other Cases: Colborn and Kangas

It may be helpful to reflect on the ways these same sorts of decisions have played out in other contexts. As chapter 1 briefly discussed, Theo Colborn faced decisions similar to those that James Hansen and his colleagues confronted. Recall that she was at the forefront of recognizing that a number of chemicals generate toxic effects by interfering with the hormonal system. As a result, these chemicals can cause a variety of health problems at surprisingly low dose levels. Nevertheless, when she and her colleagues wrote their book, *Our Stolen Future*, in the mid-1990s, the evidence that endocrine-disrupting chemicals were causing harm to humans was fairly speculative in comparison to the evidence that effects were occurring in animals. Colborn still concluded that it was worth drawing public attention to the fact that effects on human health were indeed plausible. As in the case of Hansen, other scientists thought she went too far. An editorial in the well-respected journal *Environmental Health Perspectives* asserted that "Readers ... should be mindful ... and recognize that *Our Stolen Future* is neither a balanced nor an objective presentation of the scientific evidence on ... whether exposure to endocrine-disrupting environmental chemicals is causing significant increases in human disease."[9]

Let us consider one more case in which scientists faced similar questions. This case involved disputes in the 1980s about the extent to which deforestation—especially in tropical regions—results in species extinctions. An ecologist named Patrick Kangas gave a presentation at the Fourth International Congress of Ecology in 1986, where he argued that the extinction rates caused by deforestation might be lower than researchers previously thought. His claims generated a vigorous dispute over the following two years in the pages of the *Bulletin of the Ecological Society of America*.

8. Weart 2014.
9. Lucier and Hook 1996, 350.

An influential ecologist, Reed Noss, criticized Kangas for drawing questionable conclusions that could harm society, insofar as developers could appeal to Kangas's results to promote deforestation. Other ecologists argued that it was inappropriate for Noss to try to censor Kangas. They worried that he was trying to make sure ecologists' results "come down on the 'right' side of ecologically important issues."[10] These ecologists worried that Noss's fear of disseminating results that might harm conservation efforts could harm science's reputation for objectivity. For his part, Kangas insisted that he was not supporting deforestation, and Noss maintained that he was merely trying to promote responsible science communication.

By juxtaposing the Kangas case with the Hansen and Colborn cases, we can see how difficult these issues really are. In all three cases, the scientists who spoke out with their conclusions were challenged for doing so. But as Noss's critics emphasized, science can be weakened when scientists are muzzled as well as when they speak too aggressively. When scientists honestly believe that the available evidence supports a particular conclusion, there is value in allowing them to speak out. Objectivity can be threatened when scientists are forced to withhold their judgment as well as when they present their conclusions too confidently. The Kangas case also illustrates that scientists have to consider how their responsibilities vary depending on the contexts in which they find themselves. Noss's criticisms of Kangas would have been much more compelling if Kangas had been announcing his findings in front of Congress the way Hansen did. *Responsibility to share info varies depending on info holder's context.*

Reasoning about the Cases

We have seen that scientists seeking to communicate responsibly in these sorts of cases—where there is a good deal of uncertainty and important ramifications for the public—face a challenging conflict between different values. On one hand, if they attempt to be very cautious and to avoid making any questionable interpretations of the available data, they run the risk of failing to warn society of important threats. They also risk leaving policymakers confused and unable to make informed decisions. On the other hand, if they take greater liberties in interpreting data and drawing conclusions, they run the risk of losing their objectivity and drawing false conclusions that are influenced by their personal values. Given that scientists typically aim both to benefit society and to be objective, it is not easy to navigate this conflict. One could attempt to dissolve the problem by arguing that the best way for scientists to benefit society is by remaining as objective as possible. Unfortunately,

10. Quoted in Shrader-Frechette 1996, 82.

Q/A
answer

this attempt to sidestep the conflict is a bit too simple. Even if one were to conclude that objectivity should indeed be the ultimate value for scientists, one would still be making a genuine sacrifice by prioritizing objectivity over all other values. Society might benefit by being able to trust the scientific community to be careful and objective, but it would also lose the opportunity to receive straightforward, easily understood warnings from scientists as soon as threats begin to emerge.

Faced with this apparent conflict between the values of serving society and promoting objectivity, many scientists are likely to lean toward interpreting the available information cautiously and avoiding any controversial interpretations. Philosopher Carl Cranor refers to this strategy as the "clean-hands-science, dirty-hands-public-policy" approach, and it reflects the scientific community's traditional emphasis on prioritizing objectivity. The obvious strength of this strategy is that it enables scientists to maintain their reputation for providing trustworthy information. By avoiding any controversial interpretations of the available evidence, they are better able to stick to claims that everyone can agree on. But this approach also has weaknesses. Perhaps the most obvious difficulty, as Cranor himself points out, is that it can leave decision-makers confused and unable to make informed choices. It is often very difficult for policymakers, judges, or ordinary citizens to understand scientific evidence when it is given very little interpretation. For example, if Colborn merely described the studies that had been performed on endocrine disruptors and did not elaborate on their potential implications for human health, members of the public might not be able to "connect the dots" by themselves. They would be left wondering what these studies might mean.

Q/A
answer

One might compare this situation to that of an auto mechanic who provides a very complex description of the problems with a customer's transmission. One can imagine the customer asking the mechanic to cut through all the confusing explanations and state whether the transmission is likely to fail in the next few months. A mechanic who adopted the clean-hands-science approach would refuse to draw such a straightforward conclusion for the customer. This example clearly illustrates the weaknesses of this approach. While it promotes objectivity on the part of the auto mechanic or the scientist, it seriously hampers decision-makers.

Given these weaknesses of clean-hands-science, researchers might conclude in some cases that it is better to draw the conclusion that they find to be best supported by the available evidence and to communicate that conclusion as clearly and boldly as possible. This seems to be the approach adopted by Hansen in the case of climate change, and we have already seen the concerns raised about it by other scientists. To contrast this strategy with that of clean-hands-science, we can refer to it as the "advocacy" approach. In reality, of course, scientists typically do not act in a manner that perfectly fits the clean-hands-science or the advocacy approach; their communication strategies generally fall on a continuum between the extremes.

Philosopher Kristin Shrader-Frechette suggests a potential approach for lessening the conflict between benefiting society and promoting objectivity. She argues that if we want the concept of objectivity to be helpful in the real world, we need to broaden it somewhat. It does little good to expect scientists to provide unbiased information to the public if their pronouncements are completely misinterpreted or misused by those who receive them. Thus, Shrader-Frechette suggests that objectivity encompasses not only the effort to provide information in an unbiased manner but also the effort to prevent misunderstandings of that information. The significance of Shrader-Frechette's suggestion is that she opens the door for scientists to provide additional interpretation of their findings while maintaining their objectivity, as long as they couch their interpretations in appropriate caveats and clarifications.

Consider how Shrader-Frechette's broadened conception of objectivity applies to the Kangas case. One can regard Noss's criticism of Kangas as the complaint that Kangas did not adequately take into account the ways that his presentation could be misused by interest groups who wanted to promote deforestation. On this view, Noss was not calling for Kangas to sacrifice objectivity for the sake of promoting social values. Rather, he was calling for Kangas to be more objective by adding appropriate qualifications and caveats to his presentation. For example, Kangas could have taken additional steps to acknowledge the limitations of his analysis, to clarify that other interpretations were possible, and to warn that his results should not be taken as a license to engage in increased deforestation.

Shrader-Frechette would presumably suggest that the apparent value conflicts faced by Hansen and Colborn are also not as severe as they initially appear. By adopting her broadened conception of objectivity, Hansen and Colborn could feel comfortable drawing public attention to the issues that concerned them as long as they acknowledged the limitations in the available evidence. In fact, this seems to be precisely what Hansen's fellow scientists wanted him to do. As Richard Kerr noted in his piece in *Science*, "what really bothers them [i.e., Hansen's scientific colleagues] is not that they believe Hansen is demonstrably wrong, but that he fails to hedge his conclusions with the appropriate qualifiers that reflect the imprecise science of climate modeling."[11] Perhaps this approach of hedging conclusions could be regarded as a modified form of the clean-hands-science approach that Cranor discussed. Rather than avoiding drawing conclusions, scientists could instead draw conclusions while carefully acknowledging the uncertainties and limitations associated with them.

11. Kerr 1989, 1041.

Colborn and her co-authors did in fact make efforts to hedge some of their significant claims in *Our Stolen Future*. For example, consider their cautions about the evidence that endocrine-disrupting chemicals could contribute to breast cancer: "Because of our poor understanding of what causes breast cancer and significant uncertainties about exposure, it may take some time to satisfactorily test the hypothesis and discover whether synthetic chemicals are contributing to rising breast cancer rates."[12] Their overall summary of the evidence regarding endocrine disruptors also highlights significant uncertainties: "At the moment, there are many provocative questions and few definitive answers, but the potential disruption to individuals and society is so serious that these questions bear exploration."[13]

It would be wonderful if we could stop here and conclude that the apparent conflicts between the values of promoting objectivity and serving society could be resolved by providing appropriate hedges and acknowledgments of uncertainty. Unfortunately, this is not as simple a solution as it initially appears. Consider Kerr's observations about the climate-change case:

> Experts had been hemming and hawing for a decade on the likely magnitude of the problem, and hardly anyone had listened. Then came Hansen. Now greenhouse scientists have the attention they have wanted but for reasons they think unsound.[14]

In other words, Kerr observed that when scientists tried to be responsible and carefully couch their findings in caveats and discussions of uncertainty, they lost the public attention that their claims merited. Therefore, Hansen decided that sometimes it is more socially responsible to make somewhat bold, relatively straightforward claims in an effort to attract adequate public attention.

Another difficulty is that even if scientists take steps to acknowledge the uncertainty associated with their conclusions, these clarifications can easily be ignored as they pass through the media and on to the general public. For example, experts in risk perception have highlighted the phenomenon of "availability cascades," which involve chain reactions whereby people become increasingly convinced of the seriousness of risks as information moves through society. For example, in February 1989 the TV show *60 Minutes* aired a broadcast in which the Natural Resources Defense Council presented evidence that Alar, a growth regulator used on apples and other food products, appeared to be carcinogenic. This resulted in plummeting sales of apple juice and apple sauce over the coming months, and apple growers sued the TV show for tens of millions of dollars in lost profits. Following the broadcast,

12. Colborn et al. 1996, 185.
13. Colborn et al. 1996, 231.
14. Kerr 1989, 1041.

there was a great deal of controversy over the scientific legitimacy of the NRDC report. In hindsight, it does appear that Alar is carcinogenic, but the public reaction was probably out of proportion to the seriousness of the risk. In other cases, such as public suspicion of vaccines, there can be serious public-health consequences as a result of people's tendency to take preliminary bits of scientific information (or misinformation) and blow them out of proportion.

The take-home lesson of this discussion is that it is crucial for scientists to reason carefully about their values if they are to achieve the goal of communicating their findings responsibly. When they collect evidence in support of a potential threat to human health or the environment (or when they find evidence that an alleged threat is less severe than it previously appeared), there are no easy answers about what to do (see table 5.1). On one hand, if they suspend their judgment or merely present a great deal of complicated evidence (in keeping with the clean-hands-science approach), they risk causing harm to society. On the other hand, if they draw straightforward, easily understood conclusions (in keeping with the advocacy approach), they risk sacrificing their objectivity or creating a massive scare that can have unintended consequences of its own. Scientists can partially address these concerns by acknowledging the uncertainty of their findings (in keeping with the modified clean-hands-science approach), but this can dilute the effectiveness of their message, and these cautionary notes can easily be lost in subsequent public discussions about their work.

There do not appear to be any easy, universal solutions to these challenges. Scientists need to consider, in the particular circumstances in which they find

Table 5.1 AN OVERVIEW OF SOME MAJOR APPROACHES TO COMMUNICATING ABOUT UNCERTAINTY THAT LIE ON A CONTINUUM BETWEEN VERY CAUTIOUS AND VERY BOLD

Approach	Primary Value Served	Distinctive Characteristics	Concerns or Disadvantages
Clean-hands-science	Objectivity	Avoiding interpretations and conclusions; sticking to the data	Can leave decision-makers confused
Modified clean-hands-science	A broadened conception of objectivity	Interpreting the evidence while hedging conclusions and clarifying uncertainties	Clarifications can be misunderstood, ignored, or generate confusion
Advocacy	Serving society	Interpreting the evidence to arrive at clear, easily understood conclusions	Can damage trust in the reliability and objectivity of science

themselves, whether there is more to be gained by drawing bold conclusions and communicating them to the public or by withholding their judgment for the time being. They also have to consider whether there are compromise measures that they can take to communicate their findings while minimizing the potential for people to misinterpret them. This requires weighing the seriousness of the threats that they are studying, the potential for people to misinterpret their conclusions, and the feasibility of preventing those misunderstandings. It also requires considering how much evidence to demand before drawing a conclusion, as we will see in the next section.

CHOOSING STANDARDS OF EVIDENCE

One aspect of these debates about how confidently to present uncertain scientific information is the question of how much evidence scientists should demand before drawing conclusions about socially relevant topics. Philosopher Heather Douglas has explored this issue in great detail, and she has used scientific studies of dioxin as one of her prominent examples. Dioxin has developed a reputation as "one of the most toxic substances known to humans" because it is so carcinogenic.[15] It also has the sorts of endocrine-disrupting properties that Theo Colborn identified. In addition to its toxicity, it is particularly worrisome because it lasts for an extended period of time in the environment and can build up in animal tissues. Dioxin has been associated with several chemical disasters. In one particularly infamous case, hundreds of thousands of US troops and Vietnamese citizens were exposed to dioxin that was present as a contaminant in Agent Orange, an herbicide used widely to kill crops and clear jungle during the Vietnam War. In another famous case in 1976, a chemical factory in Seveso, Italy, accidentally exposed thousands of people to a cloud of dioxin-containing chemicals, and in yet another case the town of Times Beach, Missouri, was abandoned in the early 1980s because of dioxin contamination. Apart from these particularly severe cases, dioxins are released into the environment on a regular basis as a by-product of manufacturing processes and waste incineration.

Because of concerns about the toxicity of dioxin and other chlorinated organic compounds, chapter 3 pointed out that some groups have proposed that chemical companies should try to phase chlorine out of their manufacturing processes. The chlorine industry has fought vigorously against these proposals. They have funneled tens of millions of dollars into public relations and lobbying efforts that have influenced the very highest levels of government. After the Agent Orange scare, the US Centers for Disease Control (CDC)

15. Beder 2000, 151.

began a study of its health effects but ultimately abandoned the study. A subsequent congressional hearing found that the study was "flawed and perhaps designed to fail," perhaps because of pressure from the Department of Defense and the chemical industry.[16] The chlorine industry was especially effective at influencing the Environmental Protection Agency (EPA). The chief of the EPA, Anne Gorsuch Burford, resigned in 1983 because of her questionable handling of dioxin-contaminated sites in the United States, and her successor, John Hernandez, also resigned because of inappropriate industry influences on an EPA report concerning dioxin.

The chemical industry also worked to influence the science produced on the toxicity of dioxin. According to scholar Sharon Beder, chemical companies like Monsanto and BASF produced a number of studies using questionable methodologies that were designed to minimize any findings of harmful effects. Some particularly influential articles published in journals like *Scientific American, Science,* and the *Journal of the American Medical Association* were later found in a court case to contain apparent falsifications. When a prominent environmentalist attempted to publicize these findings, industry scientists sued him for libel. The industry organized scientific conferences to help disseminate their preferred views on dioxin, and they reanalyzed previous studies in an effort to obtain more favorable results.

Douglas's Argument from Inductive Risk

Some of these industry activities appear to be clearly problematic, insofar as they involved outright falsification of research results. In other cases, however, the industry may have been adopting value-laden assumptions in a manner that could be appropriate, as long as those assumptions were made sufficiently transparent and subjected to adequate critical scrutiny. In order to think more carefully about the differences between appropriate and inappropriate influences of values in the dioxin case, Douglas distinguishes between "direct" and "indirect" ways that values could influence scientific reasoning. Values influence scientists directly when values are treated as if they are a form of evidence. For example, if environmentally oriented scientists decided to adopt the conclusion that dioxin is harmful whether or not the evidence supported their position, they would apparently be allowing values to influence them directly. Similarly, if the scientists who worked for Monsanto and BASF were implicitly or explicitly pressured by their employers to draw the conclusion that dioxin has no harmful effects regardless of the available evidence, they would also seem to be influenced by values operating in a direct

16. Shrader-Frechette 2007, 52.

way. This would be a form of the problem of wishful thinking, as discussed in chapter 1.

In contrast, scientists are influenced indirectly by values when they adjust the amount of evidence that they demand in order to draw conclusions. For example, if a chemical were suspected of being deadly, scientists might demand less evidence before deciding to warn the public about its likely harmfulness than if the chemical were merely suspected of causing skin problems. Douglas argues that it is inappropriate for values to influence scientists directly, but she thinks that the indirect role for values is crucial for scientists to take into account. This is often called the argument from *inductive risk*, meaning that when scientists face the risk of drawing erroneous conclusions, values have a role to play in choosing how much evidence they should demand.

According to Douglas, scientists are like everybody else insofar as they have a responsibility not to cause harm to other people in a negligent or reckless fashion. We would all consider it ethically unacceptable for someone to drive down a busy downtown street while constantly texting. We expect people to take reasonable steps to foresee the potential consequences of their actions and to avoid doing things that could plausibly cause harm—unless those harms are outweighed by other benefits. Douglas argues that scientists have these same responsibilities, but they manifest themselves in special ways. One of the distinctive activities that scientists perform on behalf of society is to make authoritative claims that can guide decision-makers. Unfortunately, scientists always face the risk of drawing erroneous conclusions. Therefore, Douglas insists that scientists have ethical responsibilities to consider the potential consequences of being wrong when deciding how much evidence they should demand in order to draw conclusions.

Scientists face these sorts of decision all the time when they perform statistical tests. For example, to determine whether a chemical like dioxin causes cancer, scientists typically compare its effects on two different groups of rats. One group is exposed to the chemical and the other is not. They then compare the number of tumors in the exposed group versus the "control" group. If the exposed group has more rats with tumors, the scientists can use statistical techniques to determine how likely it is that the exposed group would have ended up with that many more tumors purely as a matter of chance. It is very important to perform this sort of analysis because the group of rats exposed to the experimental chemical could end up with more tumors than the controls merely because of random variations in the health of the animals rather than because the chemical caused the tumors. Assuming that the statistical analysis indicates that the likelihood of the exposed group's randomly ending up with so many more tumors than the control group is sufficiently low—say, 5%—the scientists conclude that dioxin caused the tumors. When scientists make the decision to accept a hypothesis only when the likelihood of obtaining the results by chance is less than 5%, it is said that they are employing a "95% statistical significance level."

Douglas and others have pointed out that the choice of 95% as the statistical significance level is a very important value judgment. There is no reason why scientists could not choose a more permissive standard, such as 90%, or a more restrictive standard, such as 99%. For example, to choose a 99% statistical significance level means that the scientists would not accept a hypothesis unless their experimental results would have occurred only 1% of the time by chance. In some areas of science, it is indeed common to choose a statistical significance level of 99% or higher.

A similar situation occurs when scientists try to weigh the evidence from multiple studies to decide what conclusions they can reasonably draw. For example, they might have several animal studies, with some of them indicating that dioxin is harmful and others failing to do so. They might also have some information about the effects of dioxin on human cell cultures and a limited body of epidemiological data about the effects of dioxin on human populations. Frequently, this combination of evidence ends up being somewhat ambiguous. Different scientists can arrive at different conclusions about whether a chemical like dioxin is harmful, depending on how much evidence they demand in order to draw that conclusion. But scientists frequently do not make their standards of evidence explicit; instead, they are often subconsciously influenced by their values and inclinations.

Douglas argues that scientists have ethical responsibilities to consider the consequences of being wrong when they choose statistical significance levels and decide how to weigh evidence. For example, if scientists make a *false positive error* in a case like this one—meaning that they falsely claim that dioxin is harmful—it would create regulatory costs for chemical manufacturers. In contrast, if the scientists make a *false negative error*—meaning that they falsely claim that dioxin is not harmful—it would potentially cause people to be exposed to harmful levels of dioxin, thereby resulting in human suffering and healthcare costs. Traditionally, scientists have chosen statistical significance levels with the goal of avoiding false positive errors because they do not want to make new scientific claims that end up being false. But one could alter these approaches somewhat. For example, if one were very concerned about promoting the economic success of the chemical industry, one could raise the statistical significance level in studies even higher than 95% so that it would be harder to declare dioxin to be harmful. One could also demand that dioxin be shown to be harmful in multiple studies, including in human populations and not just in animals. Alternatively, if one thought that public health should be valued particularly highly, one might call for lowering the statistical significance level for individual studies, or one might conclude that dioxin is harmful based on only one or two studies that appear to provide evidence of harm.

Douglas's argument becomes all the more intriguing when one considers some of the other situations in which scientists have to choose standards of evidence. For example, Douglas points out that a very significant study was performed on rats during the 1970s, and tissues taken from the rats were mounted on slides so that toxicologists could evaluate them. In 1978, a group of scientists from Dow Chemical Company published an evaluation of the slides, identifying the number of rats that appeared to have benign or malignant tumors. In 1980, the slides were reanalyzed by the EPA, and in 1990, the slides were examined yet again by an independent company hired by industry. Because it is often difficult to tell which slides show evidence of tumors, the scientists involved in these three analyses came to different conclusions about how many slides had tissue with tumors and which ones were benign or malignant. Douglas suggests that when scientists are forced to make these sorts of decisions about how to interpret ambiguous information, they have to decide how much evidence to demand in order to draw particular interpretations.

For example, if scientists were particularly concerned about public health and wanted to avoid making false negative errors (i.e., falsely declaring dioxin to be harmless), they might decide that ambiguous slides should be classified as having tissue with tumors. In other words, they would be concluding, on the basis of limited evidence, that tumors were present. On the other hand, if scientists were particularly concerned about protecting the economic interests of those making or using industrial chemicals and therefore wanted to avoid making false positive errors (i.e., falsely declaring dioxin to be harmful), they might decide that ambiguous slides should be classified as having tissue without tumors. In other words, they would be demanding much more evidence before concluding that tumors were present.

In a similar way, one could consider how scientists might need to choose their standards of evidence when making many of the other choices discussed in this book. For example, as we saw in chapter 3, scientists have to make assumptions about how to extrapolate from the effects observed in animals at relatively high doses to the effects that might occur at much lower doses. When they have limited evidence available to help them make these assumptions, they have to decide how much evidence to demand before accepting one assumption rather than another. According to Douglas, this is another situation in which values can legitimately affect scientists' standards of evidence; they can alter the amount of evidence they require for making one assumption rather than another, depending on the likely consequences of making those assumptions. Table 5.2 provides an overview of some of the situations discussed in this chapter where values could influence the standards of evidence that scientists demand.

Table 5.2 ROLES FOR VALUES IN SETTING STANDARDS OF EVIDENCE

Examples of Situations in Which Values Can Influence Scientists' Standards of Evidence	Illustrations of These Value Influences in the Dioxin Case
(1) Setting statistical significance levels or making other statistical choices	(1) Choosing a level of significance for interpreting rat tumor studies
(2) Deciding how to weigh evidence	(2) Deciding whether animal studies are adequate for concluding that dioxin is harmful
(3) Interpreting data	(3) Deciding whether rat tissue slides display evidence of benign or malignant tumors
(4) Making assumptions	(4) Choosing how to extrapolate from high-dose effects to low-dose effects

Debating Douglas's Argument

Not everyone is completely comfortable with Douglas's suggestion that values can legitimately influence the standards of evidence that scientists demand. Recalling the clean-hands-science approach discussed earlier in this chapter, one might argue that scientists should try to avoid drawing conclusions that require setting standards of evidence. Proponents of this approach would suggest that scientists should merely present the available evidence to policymakers and let them decide what standards of evidence to demand in order to draw conclusions from the evidence. Of course, we have already seen that one of the disadvantages of this approach is that policymakers might get confused by all the evidence and end up making poor decisions about how to interpret it. Another problem is that the evidence itself can already be influenced by value-laden decisions; for example, Douglas's example of the rat liver slides shows that what might seem to be straightforward evidence (e.g., data about the number of slides that exhibit tumors) is actually the result of value-laden choices (namely, about how to interpret ambiguous slides).

To clarify the debates over Douglas's approach, it is helpful to make some additional distinctions about how it could be implemented. First, one might want to distinguish cases where individual scientists make decisions about standards of evidence from cases where larger groups of scientists (such as teams, networks, or societies) make these decisions together. In general, it seems easier to justify having groups of scientists make these value-laden decisions than to have individual scientists doing so. For example, scientific communities typically have conventional standards that have grown up around their standards of evidence, such as the demand for a 95% statistical significance level. Of course, these community standards are not value-free. In the case of toxicology studies, for instance, scientists could be more protective of public health if they lowered the significance level, and they could assist the chemical industry if they raised the significance level. Nevertheless, one might conclude that more harm than good would come from having individual toxicologists choose different significance levels in different cases. Perhaps the best course of action is for scientific communities to acknowledge that their typical statistical approaches are value-laden and to discuss periodically whether they strike an appropriate balance between false positives and false negatives.

Similarly, one might feel uncomfortable about having individual toxicologists interpret rat slides differently depending on their idiosyncratic views about the best standards of evidence to use for deciding whether tumors are present. This subjectivity could be alleviated by developing strategies for constraining scientists so that they have less freedom to alter their standards of evidence. One strategy for doing so would be to create community guidelines that specify how ambiguous slides should be interpreted. But once again, it is crucial to recognize that these guidelines would not eliminate the role of value judgments; they would merely shift the value judgments from the individual scientists to the community that creates the guidelines. Their decisions about how to set the guidelines would still have significant ramifications for society, and thus it would presumably be best for them to consider the social effects of setting higher or lower standards of evidence for declaring tumors to be present.

Developing community standards for setting statistical significance levels and community guidelines for interpreting evidence exemplifies an important theme in contemporary science. Researchers are gaining an increasing appreciation for the benefits of working as part of scientific teams and networks, as well as the importance of creating public repositories of data and standards for data collection so that these teams and networks can share materials. By working in groups, scientists are able to address more complicated and important social problems. In addition, diverse groups of scientists can also do a better job of addressing value judgments (as we will discuss again in chapter 7). Rather than depending on individual scientists to make decisions about how to interpret uncertain information, society is typically better served when groups of scientists reflect on how to make these value judgments. For example, while the individual judgments of prominent scientists like Hansen (and his research team) can be helpful in some cases, governments are typically better off depending on reports from scientific bodies like the US National Academy of Sciences. In the case of climate change, this is one of the reasons for the creation of the Intergovernmental Panel on Climate Change (IPCC), which synthesizes the combined expertise and perspectives of climate scientists from around the world in order to characterize the current state of knowledge.

Another distinction that can be helpful when thinking through the legitimacy of Douglas's approach is to consider whether a particular body of science is being produced primarily for scientists or for other social groups. When science is being produced primarily for other scientists, it might seem somewhat questionable to appeal to social values when setting standards of evidence. In contrast, philosopher Carl Cranor has pointed out that when scientists are focusing primarily on other groups, such as when they provide testimony in legal settings, it is important to understand the standards of evidence expected by those groups. For example, if a citizen in the United States sues a chemical company and accuses it of injuring him as a result of exposure to a toxic chemical, the citizen is required only to provide a "preponderance of evidence" in order to win the case. In the US legal system, a preponderance of evidence means that there is at least a 50% chance that the citizen is correct. Thus, the standards of evidence that scientists typically expect from each other in order to establish that a chemical is harmful appear to be much higher than the standards of evidence needed to establish that a chemical causes harm in a lawsuit. Cranor contends that scientists cause a great deal of harm when they do not understand the standards of evidence expected in other contexts, such as the legal system.

Similarly, as considered in chapter 3, when citizens are concerned about their exposure to potentially hazardous substances, they are likely to demand less evidence than scientists typically expect from each other. For example, citizens might very well be willing to take action to minimize their exposures to a chemical on the basis of preliminary evidence that it might cause harm. With this in mind, it is important for scientists working in these social contexts to consider how their conclusions might differ if they chose less demanding standards of evidence than they typically do when working with their fellow scientists.

When deciding whether it is justifiable for scientists to incorporate values in their decisions about setting standards of evidence, it is also important to consider whether they meet the additional conditions discussed in the first chapter of this book. For example, it seems much more justifiable for scientists to incorporate values in these decisions—even if they are individual scientists making the decisions for themselves—as long as they are very transparent about what standards of evidence they are demanding and why. Similarly, when scientists are generating information that will be used in applied contexts, such as regulatory policy or legal proceedings, it is particularly important to consult with other stakeholders about the appropriate standards of evidence to demand. Thus, in many contexts the important question is not whether scientists should allow values to influence their standards of evidence. Whether or not scientists consciously allow values to influence them, it is virtually inevitable that their standards of evidence will be value-laden, in the sense that they will serve some social values rather than others.

In sum, it is probably best for scientists to learn to be more explicit about their standards of evidence and their reasons for setting them in particular ways.

GENERATING UNCERTAINTY

In the beginning of this chapter, we saw that James Hansen was widely criticized by those who thought he was too quick to publicize his conclusions about the status of climate science in the late 1980s. Some of this criticism stemmed from legitimate disagreements among respected climate scientists about how to weigh different values, such as promoting scientific objectivity versus serving the public good. But a significant body of criticism stemmed from an organized climate-change-denial movement that blossomed in the late 1980s and that continued even after the evidence in support of climate change became much more compelling. Historians like Naomi Oreskes and Erik Conway and sociologists like Aaron McCright and Robert Brulle have studied how powerful interest groups "manufactured" uncertainty about climate science. This manufactured uncertainty was often not the sort of uncertainty that scientists talk about in their papers, but rather a public perception of uncertainty or controversy that helped to prevent policy action to address climate change.

In their book *Merchants of Doubt*, Oreskes and Conway trace the story of a group of scientists who tried to promote doubt about a series of environmental and public health threats during the latter decades of the twentieth century. The scientists at the center of their story were renowned physicists who helped to launch a right-wing think tank, the Marshall Institute, which played a leading role in the late 1980s in discouraging the George H. W. Bush administration from taking action to address climate change. Typical of these scientists was Fred Seitz. He served as president of both Rockefeller University and the US National Academy of Sciences and received numerous awards, including the National Medal of Science. He was also hired by the tobacco company R. J. Reynolds to run a major grants program for them. In addition, he served as an advisor for The Advancement of Sound Science Coalition (TASSC), an organization created on behalf of another tobacco company, Philip Morris, in an effort to deny the harmfulness of tobacco smoke, pesticides, fast food, and other public-health threats.

Seitz and other conservative physicists were deeply concerned during the 1980s by the criticism that left-wing scientists and organizations were lobbing at Ronald Reagan's Star Wars national defense initiative. In response, they founded the Marshall Institute. Fred Singer, another physicist affiliated with the Institute, had previously worked on campaigns to challenge acid rain and the ozone hole and had also worked with TASSC and other public relations (PR) efforts on behalf of industry. Another Institute leader, Bill Nierenberg, had been the leader of a panel that minimized the significance of acid rain on behalf of the Reagan White House.

Even though the Marshall Institute was initially founded in support of the Star Wars Initiative, it quickly turned its focus toward challenging the emerging evidence for climate change. After Hansen's testimony before Congress, Bill Nierenberg orchestrated the creation of an influential report on this topic. According to Robert Jastrow, another leader of the Marshall Institute, "It is generally considered in the scientific community that the Marshall report was responsible for the [George H. W. Bush] Administration's opposition to carbon taxes and restrictions on fossil fuel consumption."[17] Prominent climate scientist Stephen Schneider lamented that White House chief of staff John Sununu responded to concerns about climate change by holding up the report "like a cross to a vampire."[18]

Members of the Marshall Institute also attacked Ben Santer, a rising young star in the climate science community. The Intergovernmental Panel on Climate Change (IPCC) was created to be an international organization that synthesized the best available information about climate change. In 1994, Santer was selected to be the lead author for one of the chapters in the next IPCC report. He later won a MacArthur "genius" award and a number of other distinctions. Nevertheless, when the report came out in 1996, he was attacked by Marshall Institute scientists for allegedly making inappropriate last-minute changes to his chapter. (The changes were actually made according to IPCC protocol, based on last-minute comments from peer reviewers.) Fred Seitz wrote a highly publicized op-ed piece for the *Wall Street Journal* that accused Santer of fraud, and other critics tried to get him fired from his position at the Department of Energy. These attacks were so inappropriate that the American Meteorological Society actually published "An Open Letter to Ben Santer" that defended him against the accusations published in the *Wall Street Journal*.

When one encounters a story like this, it is natural to wonder how these renowned scientists ended up engaging in such questionable activities. According to Oreskes and Conway, they engaged in fairly crude misrepresentations of the available evidence in order to undermine the conclusions of the world's best climate scientists. Unfortunately, the Marshall Institute scientists were following in the footsteps of a growing community of think tanks, front groups, product-defense companies, and PR firms that specialized in casting doubt on science that challenged the interests of powerful corporations. In the middle of the twentieth century, the big tobacco companies perfected a number of strategies for fomenting doubt about science

17. Oreskes and Conway 2010, 190.
18. Oreskes and Conway 2010, 186.

that clashed with their interests. An internal tobacco industry document even stated, "Doubt is our product." Some of the strategies employed by the tobacco companies included selectively funding research projects designed to distract attention from tobacco's harmful effects, withholding worrisome research findings, challenging and reanalyzing studies that purported to show harm, attacking opposing scientists, and developing PR and lobbying campaigns to spread their preferred messages.

Big Tobacco is hardly alone in its use of these strategies. We have already seen in this chapter that the chlorine industry engaged in similar efforts. Even earlier in the twentieth century, the lead and asbestos industries also attempted to suppress unwelcome information. In his book, *Doubt is Their Product*, former government scientist David Michaels recounts how both big corporations and government agencies tried to manipulate and hide scientific evidence about the harmful properties of chemicals like beryllium, chromium, vinyl chloride, benzene, and perchlorate. And we already saw in chapter 3 that a number of critics, including prominent medical journal editors, have highlighted the pervasive efforts of pharmaceutical companies to manipulate scientific evidence on behalf of their drugs.

Given the huge financial stakes for many fossil fuel, chemical, tobacco, and pharmaceutical companies, it is not surprising that they would do everything in their power to challenge scientific information that threatens their products. The same financial incentives, as well as political pressures, can explain why powerful government agencies like the Department of Defense would be slow to acknowledge that their activities damaged the environment or harmed human health. But this still does not explain why excellent physicists like Seitz, Nierenberg, Singer, and Jastrow would be motivated to align themselves with groups engaged in relatively crude manipulations of science. Oreskes and Conway hypothesize in *Merchants of Doubt* that the key factor for these scientists was their passion for free-market capitalism. They were deeply influenced by the Cold War and by their hatred of communism. According to Oreskes and Conway, the physicists they studied in their book appeared to regard environmental regulations as a new threat to capitalism comparable to the earlier threat of communism. Thus, they were motivated to do everything possible to challenge scientific evidence that seemed to promote government regulations. As a result, they fought against the evidence for acid rain, the ozone hole, and climate change, as well as against research that supported regulations of consumer products like tobacco.

According to social scientists, most of us display to at least some degree the tendencies of these physicists. For example, we all display confirmation bias, meaning that we tend to interpret information in ways that support our preexisting beliefs and our identities as part of social groups. This effect is especially strong when we are dealing with highly polarized emotional or political issues, such as abortion, gun control, genetically modified foods, evolutionary theory, climate change, vaccines, nuclear power, or hydraulic fracturing. In fact, social scientists have found evidence that in at least some cases, providing people with more scientific information actually increases the polarization of people's views rather than decreasing it. In other words, rather than allowing new information to alter their preexisting views, people filter the information so that they can develop more sophisticated ways to defend the views they already hold. Thus, deeply held values can actually promote scientific uncertainty, in the sense that people will work very hard (both consciously and subconsciously) to challenge scientific evidence that conflicts with their values.

Alleviating Value-Generated Uncertainty

Based on these findings, science-policy experts like Roger Pielke Jr. and Dan Sarewitz have argued that it is very difficult to use scientific information to resolve highly politicized controversies. It is extremely tempting for scientists to "beat people over the head" with scientific evidence in an effort to resolve conflicts over debated issues like climate change or evolutionary theory or vaccines. But Pielke and Sarewitz suggest that this tactic is unlikely to be fruitful. Sarewitz emphasizes that in most highly debated cases, there is enough scientific complexity that opposing interest groups can find enough bits of evidence to provide at least some support for their positions. And even if the evidence for one side of a controversy is misleading or of dubious quality, as in the case of the Marshall Institute's climate-change report, an interest group can still disseminate their interpretations of the evidence widely as long as they have enough financial or political resources at their disposal.

Pielke emphasizes another reason that science frequently fails to resolve controversies. Namely, disputes that appear to be narrowly scientific are often intertwined with legitimate debates about values. In fact, people may consciously or subconsciously focus on disagreements about science as "surrogates" or "stand-ins" for disagreements that are primarily about values. For example, as we saw in the climate-change case, the motivation for physicists like Seitz and Singer to undermine the emerging consensus about climate change seems to be grounded in their desire to avoid government regulations. If they could have been convinced that effective policy actions against climate change need not destroy the free-market capitalist system that they held dear, their opposition to the evidence for climate change would probably have evaporated. Similarly, historian Mark Largent has pointed out that many of the dubious scientific claims that are made about the links between vaccines and autism are probably just smokescreens for people's general anxieties about exposing babies to so many different vaccines in such short periods of time. And in the case of genetically modified organisms (GMOs), even though opposition groups frequently appeal to safety concerns that are not well supported by existing scientific evidence, much of their underlying opposition may stem from broader social concerns about the power of big agribusiness companies and the environmental effects of many contemporary agricultural practices.

What should we make of this realization that people's deep-seated values are at the root of many of our contemporary social controversies about science? This finding can be extremely valuable for scientists and policymakers who are trying to resolve these controversies. It is undoubtedly tempting for scientists to focus on communicating their findings in more and more compelling ways. But if values are the real driving force behind many controversies, then those who aim to solve these conflicts should probably focus more on resolving value disagreements than on generating or communicating more compelling science. For example, if much of the opposition to climate change comes from those who are concerned about the future of free-market capitalism, then the best way to resolve climate-change denial may be to find ways for businesses to make money while moving toward low-carbon technologies. If climate change could be resolved via free-market technologies, most of the opposition to climate science would probably vanish.

Of course, the obvious worry about this suggestion is that it seems unrealistic. One can imagine a climate scientist replying that the reason we need to produce good, compelling climate science is because the actions needed to address climate change are not all easy and appealing. If we do not use good scientific evidence to convince people of the seriousness of the threats that they face, then they will not accept the unappealing policies that need to be chosen. In theory, this approach sounds reasonable, but in practice it may not always work very well. For example, producing and communicating more compelling scientific information has done relatively little to resolve conflicts related to climate change, evolutionary theory, vaccines, or GMOs.

There are no easy solutions to this difficulty, but at least two approaches are worth exploring further. Both solutions focus on addressing people's deep-seated values. First, as Pielke has emphasized in his own scholarship on climate change, steps can be taken to alleviate the value-related conflicts that contribute to scientific controversies. For example, Pielke suggests that policymakers should be investing massive amounts of money toward developing new renewable energy technologies, much like the Manhattan and Apollo Projects put huge investments toward nuclear and space technologies. Even though renewable energy technologies need to be cheaper and more advanced in order to address climate change effectively, we could take dramatic steps to improve these technologies. Moreover, it makes sense to pursue renewable energy for a variety of other reasons, including promoting energy independence, decreasing air pollution, creating new jobs, and developing a more balanced energy portfolio. Thus, Pielke shows that we can take deliberate steps to find and even create common ground between people's opposing values.

A second suggestion, which is compatible with the first, is to find respected thought leaders who share different people's values and who can explain why scientific conclusions that seem to be anathema to particular groups need not be as problematic as they seem. Legal scholar Dan Kahan, who has explored a number of scientific controversies, argues that one of the reasons people are loathe to accept evidence that conflicts with their values is because it threatens their identity as part of a social community. For example, hard-core political conservatives may be skittish about accepting evidence that climate change is caused by human activities because it threatens their sense of community with fellow conservatives. In response to this finding, Kahan suggests that if a trusted member of a political or social community is willing to accept controversial scientific information, it can open the door for other members of the group to do the same. For example, if a trusted religious leader explains to his congregation that their faith tradition need not conflict with evolutionary theory, this is likely to be vastly more convincing to those who are dubious about evolutionary theory than having an agnostic scientist speak to them about the scientific evidence.

Identifying Problematic Values

There is obviously much more work to be done to understand how scientific controversies can be alleviated by focusing attention on the underlying values that generate them. But in the final pages of this chapter, let us turn briefly to a slightly different question. The central focus of this book has been to argue that values have a legitimate role to play in many aspects of scientific activity. In some of the social controversies considered here, however, it is doubtful that values are playing an appropriate role. For example, the tobacco industry's efforts to suppress the evidence that tobacco causes cancer and the fossil fuel industry's efforts to reject the evidence for climate change seem completely unacceptable. Don't these cases illustrate that political and economic values ought to be excluded from science?

The influences of values in these cases do indeed appear to be unacceptable, but that does not mean that values must always be excluded from science. It also does not mean that the values of corporations are inherently problematic. As we saw in chapter 1, we need to keep thinking about how to distinguish between legitimate and illegitimate value influences. The cases in this chapter provide further support for two of the conditions that we have identified throughout the book. First, crucial methodological choices and assumptions typically need to be acknowledged transparently and honestly so that others can determine if they agree with the value implications of those choices. This seems to be a crucial difference between the Marshall Institute scientists and the climate scientists who criticized James Hansen. The climate scientists raised the legitimate concern that the evidence might not be solid enough to justify the relatively definitive conclusions that Hansen was attempting to draw. In contrast, the Marshall Institute scientists tried to smear Ben Santer with false accusations. Similarly, the tobacco industry willfully misrepresented their research throughout the twentieth century. Top executives knew that tobacco smoking was harmful and addictive, but they funded strategic research projects in an effort to distract the public from coming to these conclusions. When interest groups misrepresent and withhold evidence, or when they persist in raising objections without acknowledging that those concerns have already been addressed by the scientific community, they fall prey to the problem of wishful thinking. This is a problem no matter what the political or ideological persuasion of the interest group might be—whether it is aligned with corporations or government agencies or nongovernmental organizations.

A second condition is that the values influencing our research enterprise should adequately represent fundamental ethical principles and, when those do not settle the matter, the values of those who will be affected by the research. For example, we saw at the end of chapter 2 that a crucial worry about contemporary biomedical research is that our current research efforts—which focus primarily on generating patentable treatments for the illnesses that afflict the world's wealthiest countries—may not accord with all the values that we espouse. Similarly, it is doubtful that most people have been as eager to exonerate chemicals like lead, asbestos, dioxin, benzene, beryllium, chromium, and vinyl chloride as the companies that produce those chemicals. Therefore, it seems likely that the standards of evidence required by the industry funders for declaring these chemicals to be harmful are probably much higher than what most citizens would demand. This does not mean that the values of industry are inherently problematic, of course. In many cases, the values of companies and the values of the general public align fairly well. Industry values are likely to be less representative of the general public's interests, however, when companies are testing the public-health or environmental impacts of products that generate a great deal of money for them.

A similar situation appears to be present in research on climate change. Sociologist Robert Brulle has shown that fossil fuel companies have funneled huge amounts of money toward think tanks and researchers who will challenge the evidence for climate change. It is doubtful that their values adequately represent those of the American population as a whole because the fossil fuel industry has much more to lose than the average person from shifting our economy toward renewable sources of energy. As a result, even when these companies are not misrepresenting the evidence, they typically demand exceptionally high standards of evidence in order to accept that human actions are contributing to climate change. Thus, it is worrisome if their conclusions receive disproportionate attention because of their ability to trumpet their perspectives through the powerful think tanks and PR companies that they support.

CONCLUSIONS

This chapter focused both on the ways in which values can generate scientific uncertainty and on the ways in which values are relevant to addressing uncertainty. The final section highlighted many worrisome ways in which values can contribute to the illegitimate "manufacture" of uncertainty. For example, in his research on the funding sources for climate-change-denial organizations, Brulle identified 91 key organizations that have been promoting skepticism about human-induced climate change. He found that each year an average of more than $900 million was donated to these organizations, which include highly influential think tanks like the Cato Institute, the American Enterprise Institute, and the Heritage Foundation. According to Oreskes and Conway, 92% of the 56 "environmentally skeptical" books published in the 1990s were linked to these sorts of right-wing foundations, along with 100% of the 13 skeptical books published in the 1980s. The last section of this chapter pointed out that in addition to disputing the false scientific claims fomented by these organizations, it is at least as important to address the values underlying their skepticism. In many polarized science-policy issues, including GMOs, vaccines, climate change, and evolutionary theory, values appear to be the driving force behind the controversies. If we can find win-win solutions and trusted individuals who can defuse the value conflicts, much of the scientific opposition is likely to melt away.

Whereas the final section of the chapter highlighted the ways that deeply held values can stir up scientific uncertainty and controversy, the first two sections focused on the important roles that values play in guiding scientists' responses to uncertainty. In the first section, we saw that researchers like James Hansen, Theo Colborn, and Patrick Kangas had to make difficult choices about how confidently to communicate their findings. Most of the scientists involved in these controversies seemed to agree that values were relevant, insofar as they thought that scientists should strive to communicate their results in a socially responsible fashion. But they disagreed in specific cases about what approach was most socially responsible. Ideally, they aimed to remain as objective as possible while also drawing the public's attention to new and emerging ideas. But we saw that determining how to strike an appropriate balance between these two goals requires careful deliberation and sensitivity to the ways that the media and the general public take up scientific information.

Building on this discussion of the role of values in communicating about uncertain information, the second section of the chapter focused specifically on the role of values in setting standards of evidence for drawing conclusions. It examined Heather Douglas's argument that scientists are frequently faced with situations where they have to choose how much evidence to demand. According to Douglas, scientists should consider the social consequences of the conclusions that they are drawing when they decide how much evidence is adequate. Nevertheless, we saw that this role for values is not entirely uncontroversial. Scientists can try to evade some of these difficult decisions either by choosing the standards of evidence that their communities typically employ or by presenting relatively uncontroversial evidence and letting the recipients of information decide what conclusions to draw. But these evasive strategies typically have disadvantages or limitations of their own. Thus, determining what role values should play in responding to uncertainty is one of the most challenging issues discussed in this book, and one that deserves much more reflection. *Overall goal reitterated: What role should values play in responding to uncertainty?*

SOURCES

James Hansen's testimony is discussed in Kerr (1989), Kessler (2015), Shabecoff (1988), and Weart (2014). Further information about Theo Colborn can be found in Elliott (2011b) and Smith (2014). Carl Cranor (1990) discusses the clean-hands-science, dirty-hands-public-policy approach. The controversy between Kangas and Noss is discussed in Shrader-Frechette (1996).

Beder (2000) and Shrader-Frechette (2007) discuss industry efforts to discourage regulations of dioxin. The strengths and weaknesses of the argument from inductive risk (i.e., the notion that values should play a role in setting standards of evidence) are discussed in de Melo-Martin and Intemann (2016), Douglas (2000; 2009), Elliott (2011a; 2013a), John (2015), and Steel (2010). Wilholt (2009, 2013) discusses how scientific communities develop conventional standards of evidence. Elliott and Resnik (2014) discuss the potential for these value judgments to be subconscious and the importance of trying to make them more transparent. Douglas (2003) discusses the responsibility of scientists to consider the social consequences of their choices, and Douglas (2007) examines the potential for multiple stakeholders to help make decisions about standards of evidence.

Information about the efforts of interest groups to promote uncertainty in the case of climate change can be found in Brulle (2014), McCright and Dunlap (2010), and Oreskes and Conway (2010). Further information about efforts to generate uncertainty and to manipulate evidence in a wide variety of different cases is available in Elliott (2016b), Goldacre (2012), Michaels (2008), and Proctor (2012). Pielke (2007) and Sarewitz (2004) discuss the difficulty of using scientific information to resolve complex public policy disputes. Pielke (2010) discusses strategies for responding to climate change. Largent (2012) discusses how people appeal to science as a surrogate for broader concerns about vaccines. Dan Kahan (2010) argues that many disputes that appear to be about science are actually grounded in people's commitment to a social group, and he offers solutions similar to the ones proposed at the end of this chapter. Biddle and Leuschner (2015) offer additional perspectives on how to distinguish appropriate scientific dissent from inappropriate dissent.

CHAPTER 6

How Should We Talk about It?

As Valentine's Day approaches each year, news stories often appear about the love lives of voles and the lessons they can teach us about human relationships. Voles are small rodents that resemble mice and hamsters. Their Valentine's Day fame arose largely because of a fascinating series of experiments performed in the lab of Larry Young, a researcher at Yerkes National Primate Research Center in Atlanta, Georgia. As Thomas Insel, the director of the US National Institute of Mental Health, has pointed out, voles are an "extraordinary gift to science."[1] This is because some species of voles, unlike the vast majority of mammals, are monogamous—at least in the sense that they form bonds with their partners that typically last long after they mate. But what makes voles particularly marvelous experimental animals is that other species of voles are not monogamous. Therefore, researchers can study the differences between these species to develop hypotheses about the reasons for their differences in mating behavior.

Research on voles also provides an excellent introduction to the challenges scientists often face when deciding how to communicate information responsibly. As we will see in this chapter, scientists often have to decide how best to frame their research and which terms or categories are most appropriate for describing it. Both of the justifications provided in chapter 1 support the conclusion that values should be incorporated in these choices about how best to communicate scientific results. First, in many cases there are no perfectly "neutral," value-free ways to provide information. Therefore, scientists cannot help but support some values over others; the best they can do is to reflect on the most appropriate values to support. The second reason for incorporating values in these choices is that socially responsible science communication

---

1. Tucker 2014.

requires not only efforts to be accurate but also the pursuit of additional goals. Depending on the situation, these may include the following: (1) explaining how new research results relate to previous findings and other areas of science; (2) clarifying how scientific results could eventually affect society; (3) preventing misunderstandings about the nature of results; and (4) clarifying the potential ramifications of scientific results for people's goals, values, agendas, or worldviews.

Because these goals involve social considerations that go beyond mere scientific accuracy, values are relevant to deciding how best to achieve and prioritize them. Admittedly, these goals are not all relevant in every instance of scientific communication, but in many cases they are present and can stand in tension with one another. One could, perhaps, lessen the role for values in science communication if one decided that scientists should abandon all goals except accuracy. However, the first justification for incorporating values in science communication would still remain: there are often no value-free ways to frame and describe scientific findings. In other words, there can be multiple ways to talk about a scientific finding or problem that are all relatively accurate, but each approach subtly privileges or supports some values over others.

This chapter explores three aspects of scientific communication that generate these sorts of value-laden choices: *frames, terms and metaphors*, and *categories*. First, we will use the vole case to explore the decisions that researchers need to make about how best to *frame* their messages. Second, we will see in a wide variety of cases that value-laden choices can be very difficult to avoid when choosing *terminology and metaphors*. In many instances, researchers will end up privileging one social perspective or another no matter which terms they choose to employ, so they have to decide which perspective is most appropriate to support. Third, we will examine how values can become relevant to choosing *categories or classification schemes*. The history of racial classification provides a particularly vivid example of the challenges that arise when choosing categories, but we will see that these issues arise in other cases as well.

FRAMES

To better understand the issues that arise when deciding how to frame scientific information, let us return to the story of the voles. It turns out that male prairie voles—one of the monogamous species—have a particularly large number of receptors for a chemical called vasopressin in a particular region of their brains. Vasopressin is a hormone that regulates the constriction of blood vessels and the retention of water, but it also appears to influence social and sexual behaviors in animals. It is very similar in structure to another hormone called oxytocin, which is sometimes called the "bonding hormone" because it appears to affect people's social relationships, trust, intimacy, and maternal behavior. Larry Young and his colleagues identified the form of the gene responsible for the large number of vasopressin receptors in prairie voles and found a way to insert copies of that gene into meadow and montane voles. Their amazing finding was that these voles—which are normally promiscuous—began to exhibit the sorts of pair-bonding behaviors typically found in prairie voles. The scientists were also able to influence voles in the opposite way; they inhibited the formation of pair bonds between the voles by blocking vasopressin or oxytocin receptors in their brains.

This research launched a fascinating set of research projects to see whether differences in behavior among humans could be traced to hormones like vasopressin or oxytocin. A neuroscientist named Hasse Walum performed experiments in which he found that variations in the same gene studied by Young's lab were associated to some extent with the reported quality of bonding between human men and their partners. Another scientist, Paul Zak, found that people displayed more trusting behavior after having oxytocin squirted up their noses. (The hormone was administered through the nose in order to get it across the blood-brain barrier.) A variety of other experiments have also found associations between these hormones or their receptors and the display of social behaviors like trust, generosity, and memory among humans.

As one might imagine, the media has had a field day with this research. The headlines abound with striking claims: "Monogamy Gene Found in People," "Gene Determines Fidelity in Men," "Why Men Cheat: Study Chalks Up Promiscuous Behavior to a Single Genetic Change," "Being Human: Love—Neuroscience Reveals All," and "The Love-Rat Gene: Why Some Men Are Born to Cause Trouble and Strife." On the *NBC Nightly News*, Brian Williams reported:

> Throughout history men have come up with all sorts of excuses for behaving badly. Now it appears they have a new one. It's in their genes, apparently. This is a line of research that started with rodents called voles. Now it's being applied to humans.[2]

These messages clearly resonated with members of the public. On the *Today Show*, for example, a woman responded to the voles research by saying, "I would want to have my mate tested before I was married. And I am single and that would secure my marriage."[3] Similarly, on *The Tyra Show*, a man said that he felt like cheating was "in my blood."[4]

These sorts of claims are not unique to the news media; some scientists and physicians have made similar assertions in response to the research. Paul

2. Quoted in McKaughan and Elliott 2013, 215.
3. Quoted in McKaughan and Elliott 2013, 217.
4. McKaughan and Elliott 2013, 215.

Zak, a professor who bridges the fields of economics and neuroscience, wrote a book called *The Moral Molecule: The Source of Love and Prosperity*. Describing the ideas in his book, he claims:

> Given that humans can be at the same time rational and irrational, ruthlessly depraved and immensely kind, shamefully self-interested and completely self-less, what specifically determines which aspects of our nature will be expressed when? When do we trust and when do we remain wary? When do we give of ourselves and when do we hold back? The answer is oxytocin.[5]

Similarly, Shirah Vollmer, a professor of psychiatry and family medicine at UCLA, writes:

> This research opens the door to medication to treat infidelity. If we improve the reward of vasopressin, then we increase the likelihood of faithful marriages. It also changes the valence of fidelity. If infidelity is a genetic variant, should physicians treat it like hypertension or diabetes? On the other hand, perhaps the infidelity gene is closely linked to the charisma gene, and as such, it is part of the package of seduction.[6]

On one hand, it is helpful for scholars to present the potential implications of cutting-edge neuroscience in a manner that is easy for everyone to understand. On the other hand, these scholars are presenting the material in ways that can generate a great deal of confusion. They are giving the impression that there is a fairly straightforward relationship between particular genes and people's behavior, but claims like these are premature. Even in voles, non-genetic factors such as maternal nurturing can have very important influences on the nervous system and on vole behaviors. In humans, there are obviously a host of non-genetic factors that affect our social behavior, and therefore it is unsurprising that the correlations that Walum observed between gene sequences and partner bonding were small. Moreover, further research in voles illustrates that relationships between genes and behavior are very dependent on the context. For example, the alleged "monogamy gene" that seemed to make such a difference when inserted into montane and meadow voles is already present in numerous vole species that do not display monogamous behaviors. Further research has also suggested that the relationships that Young observed between genes and behavior in the lab may not be present in the same ways in free-living prairie voles.

5. Quoted in McKaughan 2012, 527.
6. Quoted in McKaughan and Elliott 2013, 217.

One possible response to the story of the voles would be to insist that scientists and the media simply got carried away and made sloppy claims that were unsupported by the evidence. If one adopted this view, it might seem that scientists have no need to make difficult decisions about how to communicate appropriately in this case—they just need to "stick to the data" and quit exaggerating. It is undoubtedly wise for scientists not to exaggerate their claims, but this is perhaps too simple a solution in the case of the voles. It seems genuinely valuable for scientists to inform the public about the potential ramifications of their research, even if this requires some speculation. For example, Larry Young and other scientists have worried that oxytocin could eventually be incorporated into "trust sprays" that military or political operatives could use to manipulate people's level of trust in one another. In fact, neuroscientist Antonio Damasio argues that traditional marketing techniques may already work in part by manipulating the natural release of oxytocin in people's brains. Young has called for ethicists to begin thinking proactively about these issues.

In addition to warning people about the ways research could be abused or used to harm society, it also seems valuable for researchers to clarify how their research could influence our self-conceptions and worldviews. In an opinion essay in *Nature*, Young proposed that "biologists may soon be able to reduce certain mental states associated with love to a biochemical chain of events."[7] While this claim might be a bit too strong, it does seem helpful for Young to elaborate on the ways his research could influence our understanding of human nature. We might even think that it was irresponsible if an investigator received millions of dollars in taxpayer funding and failed to elaborate on the ways his research could help us think about broader human interests and concerns. Indeed, a prominent philosopher, Patricia Churchland, has argued that this research could help us explore questions about human morality and free will, insofar as it identifies molecular processes that influence human behavior. While there is always the danger that scientists will make speculative claims that go too far, it would also impoverish society if they entirely failed to point out these potential ramifications of their research.

So, we are faced with a significant challenge that resembles the conflict between clean-hands-science and advocacy discussed in chapter 5. While chapter 5 focused on the appropriateness of drawing bold conclusions on the basis of limited evidence, however, this chapter focuses on decisions about how to frame scientific information. On one hand, if scientists simply "stick to the data" and refuse to reflect on the significance of their findings, they

7. Young 2009, 148.

risk depriving the public of valuable information. On the other hand, if they deliberately try to frame their findings in ways that connect them to broader public concerns, they risk coloring the research with their personal values and worldviews.

A frame is a way of characterizing a message or a problem or a decision situation. Scholars have described different sorts of frames. Some frames describe an identical situation in different ways, such as when one says that a new medical treatment has a 90% survival rate as opposed to saying that it has a 10% mortality rate. Other frames emphasize particular details or considerations related to a scientific finding while deemphasizing or ignoring other details or considerations. For example, research on voles and vasopressin can be framed to emphasize its relevance for human romantic relationships, to emphasize its relevance for military purposes, or to emphasize the ways in which science could potentially "reduce" everyday human experiences to chemical and biological phenomena.

A recent discussion in the journal *Science* highlights the issues at stake in scientific framing. In an opinion piece, Matthew Nisbet and Chris Mooney argued that scientists could do a better job of communicating with the public about contested issues like climate change, evolution, and embryonic stem-cell research if they were more willing to frame these topics strategically. They contend that most people do not have the time or motivation to examine in any detail the facts and arguments related to scientific issues. Therefore, people depend on frames to help them synthesize information rapidly and determine how scientific information might connect with their own needs, concerns, and values. For example, Nisbet and Mooney point out that climate change has been framed as an issue of "scientific uncertainty" and "unfair economic burden" by conservative political groups, thereby diminishing public confidence in the science. But they note that Christian evangelical leaders have partially counteracted this skepticism by framing climate change as a matter of religious morality. They also point out that efforts to highlight political interference with climate science by the Bush administration created a "public accountability" frame that also helped to counteract the uncertainty frame.

In the case of evolution and stem-cell research, Nisbet and Mooney argue that the same sort of attention to framing could help promote greater public acceptance. They point out that a "scientific uncertainty" frame, as well as a "teach-the-controversy" frame, was used by skeptics to challenge evolutionary theory. Scientists tried to respond by offering detailed scientific responses to the critics, but these efforts largely failed. Nisbet and Mooney suggest that this is a classic case where people do not have the time or energy to wade through scientific details. Instead, it would be much more valuable to offer counter-frames, such as "social progress," which bills evolution as a building block for medical advances. They argue that the same "social progress" frame, together with an "economic competitiveness" frame, can push forward public support for embryonic stem-cell research.

Not everyone agrees with Nisbet and Mooney, however. A subsequent set of letters to *Science* challenged their enthusiasm about framing scientific information in a strategic fashion. For example, Earle Holland, a long-time science communications specialist from Ohio State University, argued that Nisbet and Mooney's suggestions seemed "somewhat dishonest," and he called for scientists to "rely on their data, rather than on what 'spin' on an issue might prove more convincing."[8] In another letter, Robert Gerst insists that "science has credibility with the public precisely because the public believes that science is neutral, that it doesn't take positions or adopt particular frames."[9] Thus, he warns that efforts to frame science will undermine its credibility and destroy public trust. Even Nisbet and Mooney acknowledge that their proposal may seem "Orwellian" in the way it tries to manipulate how people think about particular bodies of science.

What are we to make of this dispute? First, it is impractical to think that scientists can communicate effectively with the public without adopting any frames; if they think their language is free of frames, they are probably just failing to recognize them. Consider the frequency with which scientists doing work on biomedical issues talk about how their research could eventually help to cure or alleviate human diseases. Or consider the physicists who emphasize how their work can satisfy our fundamental curiosity about the universe or how it could potentially give rise to new technologies. These are all frames, albeit relatively innocent ones. Given this era of tight budgets and public suspicion of government spending, it is doubtful that the scientific community could continue to drum up government funding for research without framing science in ways that highlight its significance. Thus, it is more realistic to explore the question of how science can be framed appropriately rather than trying to avoid framing altogether.

As Gerst suggested, one of the crucial problems with adopting particular frames is that they tend to support specific value orientations or worldviews and thereby threaten the neutrality of science. For example, framing embryonic stem-cell research or evolutionary theory in terms of social progress tends to elicit positive attitudes toward these areas of science. Thus, scientists might be illegitimately smuggling their own values into the presentation of scientific information. A proponent of these frames might argue that this is entirely appropriate, as long as it is true that these theories or areas of research will in fact generate social progress. But even if we assumed that social progress would definitely arise from these areas of science, there are

8. Kavanagh 2007, 1168.
9. Kavanagh 2007, 1169.

still value judgments involved in focusing people's attention on these positive features to the exclusion of other features of the science.

Strategies for Framing Information Responsibly

We have seen that it is very difficult to avoid framing scientific information, and thus values are very likely to influence the process of scientific communication. Nevertheless, scientists can still take steps to incorporate these value influences in a responsible fashion. In some cases, choices about how to frame scientific information might be so important that it would be worth engaging with social scientists or community groups to determine what frames best represent social concerns and priorities. We will discuss how this might work in the next chapter. Philosopher Daniel McKaughan and I have suggested two additional steps that scientists can take: (1) they can reflect on the major frames that are common in their area of research; and (2) when their chosen frames incorporate significant values or assumptions or weaknesses, they can try to acknowledge them. These steps are consistent with the emphasis throughout this book on the importance of making value judgments more transparent so that they can be subjected to scrutiny and deliberation.

We use the term "backtracking" to describe this process of acknowledging value judgments or concerns associated with one's chosen frames so that listeners can explore alternative interpretations. "Backtracking" is a metaphor designed to express the idea that even if scientists lead people down particular interpretive "paths" by framing information in one way rather than another, they can help people recognize the limitations of those paths and the possibility of taking other interpretive paths or frames. A number of different activities or strategies can potentially contribute to backtracking: clarifying the level of evidence in favor of different claims or frames, taking steps to prevent potential misunderstandings associated with particular frames, employing multiple frames when communicating complex ideas, acknowledging major weaknesses of one's chosen frames, and employing frames that are generally regarded as uncontroversial. Admittedly, there are major limitations to some of these strategies for backtracking. In some cases, the "damage has already been done" once scientists employ a particular frame; their listeners will always remain somewhat biased in favor of that interpretation. In addition, scientists often have little control over the way their work is framed in the media. Nevertheless, scientists can still often exert some control over how their work is presented, and even small efforts on their part to acknowledge important assumptions or limitations in their portrayal of information can go a long way to help their listeners.

Consider how this could work in the voles case. McKaughan and I have identified five frames that are particularly common. First, some scientists and media presentations emphasize a "genetic determinism" frame, according to which a particular gene or molecule is responsible for social behaviors like monogamy. A second frame, "humans are like voles," suggests that the research findings in voles provide lessons for understanding human behavior. A third frame, "triumph for reductionism," treats the research as an example of how human experiences like love can be characterized in terms of physical and chemical processes. Fourth, the "saving your relationship" frame presents the research as a potential source of lessons for improving human relationships. Fifth, a "dangers of social manipulation" frame highlights how this research could generate harmful influences on society.

By becoming aware of these common ways in which their work is framed, scientists working in the voles case could promote backtracking in a variety of simple ways. Suppose that a scientist were interviewed by a news show and discussed how the voles research helps to identify connections between genes and behavior, thereby encouraging a "genetic determinism" frame. The scientist could take pains to mention that in humans, a supposed "monogamy gene" would be only one of many different factors that influence people's behavior, and even in voles there are species that have the gene but that do not act monogamously. A small clarification like this could help the listeners realize that the genetic determinism frame has limits that could be explored further. Similarly, if a scientist felt like a media interview was focusing heavily on the "saving your relationship" frame, she could mention that the voles research could also help us to gain insights into the nature of autism and other disorders associated with social behavior. Again, this would alert listeners to the potential for thinking about the research differently.

Of course, these sorts of clarifications will not always be feasible or appropriate. For example, consider a scientist who uses a "social progress" frame to highlight the medical significance of evolutionary theory or a scientist who uses a "public health" frame to emphasize the importance of vaccinations. Would these scientists need to backtrack and discuss the "teach-the-controversy" frame about intelligent design theory or the "vaccines-cause-autism" frame? In general, the answer is no. Intelligent design theorists have not had success in launching a productive scientific research program based on their theory. The concept of intelligent design can still be explored as an important theological or philosophical topic, of course, but so far it has not generated a promising alternative to evolutionary theory. Similarly, the available evidence indicates that vaccines do not cause autism. Part of responsible backtracking is to give one's readers or listeners a sense of how well the available evidence supports different frames. Thus, it would be irresponsible to distract people from a well-supported frame by highlighting a poorly supported one. Nevertheless, we have seen that there are also cases like the voles research, where there are multiple plausible frames and they all have weaknesses. In these cases, it is particularly important for scientists to recognize the significance of their frames and their potential to support some values or worldviews over others.

It is also worth emphasizing that even though some frames should not be emphasized, this does not mean that the values underlying those frames must be excluded from science entirely. A central theme of this book is that values of all sorts (except for those that violate well-supported ethical principles) can play legitimate roles in science if the right conditions are met. Consider the religious values underlying the pursuit of intelligent design theory. We saw in chapter 4 that it can be appropriate for values to influence which theories scientists decide to develop or explore. We also saw, however, that it is crucial to be transparent about the difference between exploring a theory versus thinking that it is true. Therefore, as long as it did not waste resources that would be better spent elsewhere, it might not be problematic to investigate the potential of intelligent design theory, as long as one did not give the misleading impression that it was already well supported by the available evidence.

TERMINOLOGY AND METAPHORS

Choices about terminology constitute a second set of value-laden issues associated with the communication of scientific information. To better understand these choices, consider debates that arose in the 1990s when a number of different groups that were working to promote biodiversity and ecological restoration in the Chicago area banded together to form the Chicago Wilderness organization. This group promotes a number of initiatives, including efforts to respond to climate change, strategies for connecting children with nature, and plans for promoting green infrastructure. One of its central activities has been to preserve and restore historic ecosystems on hundreds of thousands of acres of land in southern Wisconsin, northeastern Illinois, and northwestern Indiana. For example, prairies covered much of this land in the past, and volunteers for the Chicago Wilderness organization have helped to recreate prairie ecosystems where they had been lost.

Surprisingly, these activities have generated a significant amount of controversy. Some community members challenged the Chicago Wilderness organization for engaging in destructive activities in the course of their efforts to restore ecosystems that were present in previous centuries. Paul Gobster, a social scientist with the US Forest Service, recounts some of the critics' reactions:

> People decried killing healthy trees by cutting, girdling, applying herbicide, and
> burning, and they mentioned places where numerous trees had been removed,

leaving an open, "barren" landscape. Some felt tree removal was at odds with the purpose of the forest preserves and the whole idea of "restoration."[10]

Critics worried that bird and deer populations were being harmed and that beloved recreation areas were being altered. In other words, many of the critics complained that the Chicago Wilderness organization was actually destroying parts of nature that the citizens held dear. In response, proponents of the project insisted that they needed to eliminate species that were non-native to the region in order to recreate ecosystems like prairies that were present in the past.

Some of those engaged in these conflicts over ecological restoration claim that the values embedded in scientific terminology can complicate these debates, and thus scientists need to think carefully about their choice of terms. For example, environmental scientists often refer to species that move into a new ecosystem as "alien species" or "exotic species." When they have the potential to spread and cause harm to "native species," these exotics are often called "invasive species." Environmental scientist Brendon Larson worries that a term like "invasive species" inevitably brings with it the connotations of military invasions, together with the very negative values associated with them. Larson argues that this language encourages people to adopt a "war" against invasive species. One can see how critics of Chicago Wilderness could worry that this sort of language might encourage the debatable presumption that highly aggressive activities are justified in response to non-native species.

Larson argues that these are just a few of the many environmental-science terms that involve metaphors taken from human society. These metaphors are significant because they pull values into science. Recall that a metaphor describes one item in terms of something else, thereby making an implicit comparison between those two things. For example, if we say that the night sky is studded with diamonds, we are drawing a metaphorical comparison between stars and diamonds. In his book *Metaphors for Environmental Sustainability*, Larson argues that the environmental sciences are completely saturated with metaphors. Consider just a few, out of a list of almost 100 metaphorical terms that he provides: community, competition, food chain, stability, biodiversity hot spot, biological inertia, ecological integrity, ecosystem health, natural capital, fitness, keystone species, genetic drift, and mutation. Some of these terms have become so standard in science that their metaphorical status has become almost invisible.

Larson acknowledges that many of these terms have been given precise definitions in their new scientific contexts, but he argues that they still often carry values associated with the social contexts from which they originated.

10. Gobster 1997, 33.

As mentioned earlier, terms like "invasive" or "alien" species provide good examples of this concern. Larson suggests that the values associated with these terms could promote beneficial environmental action in some cases, but in general he worries that militaristic and aggressive metaphors are not ideal for promoting long-term sustainability. They generate fearful attitudes on the part of the public, they promote sharp distinctions between humans and nature, and they exaggerate the ecological differences between exotic and native species. Moreover, references to "native" and "non-native" species create a whole other set of worrisome connotations because they encourage comparisons to social debates about human immigration.

In response to these heavily value-laden metaphors, Larson acknowledges that one could try to develop purely value-neutral terminology for talking about environmental issues. But he emphasizes that this proposal runs into the same difficulties that face those who want to eliminate value-laden scientific frames. First, it is often doubtful that purely value-neutral terminology is available. Second, such purely value-neutral terminology would probably be much less effective for communicating to the public about environmental problems. Thus, Larson calls for scientists to explore terms and metaphors that fit better with our social goals. For example, he suggests that we could shift to talking about "superabundant" or "harmful" species instead of "invasive" ones. These terms would highlight the fact that some species become too abundant and cause problems, but they would not have such militaristic overtones. Moreover, they would not draw such a sharp distinction between native species and newly introduced ones, considering that native species could also become harmful in some cases.

Larson's call for scientists to reflect on the social ramifications of their language is applicable throughout the environmental sciences. For example, philosopher Stephen Gardiner has discussed some of the linguistic issues that become significant when discussing climate change. In the 1980s, many people employed the metaphor of "the greenhouse effect," but some scholars began to think that it would be better if they used language that communicated more directly about the consequences of this effect. Thus, they shifted to the term "global warming." But this term also began to fall out of favor, in part because of the misleading impressions it could create for the public. For example, Gardiner notes that people might like the sound of a warmer globe, but they might not realize that the rapidity of these changes—not only in terms of temperature but also precipitation patterns—could have devastating effects on both the natural world and on human civilization. Using the term "climate change" does a better job of capturing this problem of change. However, science-policy expert Roger Pielke Jr. reports that some US Republican Party memos have also encouraged use of the term "climate change," but for political purposes. They seem to prefer it because it raises less worrisome connotations than "global warming" among some constituencies. Thus, responsible scientists working in this area have to weigh considerations like accuracy and understandability alongside the goals of preventing misinterpretations and not uncritically playing into the hands of political strategists.

As our ability to prevent serious climate change has become increasingly doubtful, some scientists have suggested that we should put more effort into studying climate geoengineering. This area of research explores strategies for cooling the global climate, either by removing carbon dioxide from the atmosphere or by increasing the earth's albedo (i.e., the reflection of solar radiation from the earth's surface). For example, the globe has been known to cool temporarily after volcanic eruptions because of the huge quantities of sulfur particles that are spewed into the atmosphere and that block solar radiation. Thus, some scientists have proposed that we could deliberately shoot the same sorts of particles into the atmosphere over an extended period of time in an effort to keep the planet from warming. Of course, these strategies raise a host of potential problems of their own.

There are currently major international debates about the feasibility and wisdom of pursuing geoengineering, and it is noteworthy that many of these debates start with disagreements about the best terms to use for referring to it. A 2015 report from the US National Academy of Sciences chose the term "climate intervention" rather than geoengineering, in part because the word "engineering" gives the impression of predictability and control, which is not present in proposed geoengineering schemes. But the issues do not stop there. Climate geoengineering techniques that focus on increasing the earth's albedo are often called "solar radiation management" (SRM) techniques. Once again, some critics have worried that the term "management" makes SRM seem more precise and easier to control than it really is. Thus, other scientists have proposed terms like "sunlight reflection methods" or "albedo modification" in an effort to avoid these connotations. Raymond Pierrehumbert, one of the authors of the National Academy of Sciences report, even suggested in a separate commentary that a term like "albedo hacking" would do a better job of representing the experimental and dangerous features of geoengineering techniques. As in so many of the cases discussed throughout this chapter, there do not appear to be any completely neutral terms that avoid any value connotations. Thus, responsible scientists have to consider which terms achieve the best balance of goals.

Terminology in Biology and the Social Sciences

To show how widespread the influences of values on terminology can be, the remainder of this section touches on a wide range of other scientific fields. As an example from biology, consider the voles case that launched this chapter. Scientists labeled the prairie voles and the genetically modified meadow voles as "monogamous," and this term clearly had a huge influence on public discussions about the research. On the positive side, this term is helpful because of the way it facilitates easy understanding of the behaviors displayed by the voles. But on the negative side, it can also promote confusion because the monogamous behaviors displayed by the voles are different than the behaviors that we typically associate with human monogamy. When the biologists describe voles as monogamous, they mean that they develop long-term bonds with a particular partner and engage in frequent huddling and shared care of their offspring. But these "monogamous" voles still engage in occasional sexual intercourse with individuals other than their partners, which is not the way we typically think of monogamy among humans. Thus, it would be very misleading to assume that the genetic factors that promoted "monogamy" among voles could be used reliably to prevent humans from cheating on their partners, as some media discussions seemed to suggest.

The voles case illustrates the challenges that biologists and social scientists often face when choosing between terms that are understandable and meaningful to the public versus those that are least likely to cause confusion. Philosopher John Dupré has pointed out that the same sorts of problems that arise with the term "monogamy" also arise with other terms, like "rape." The concept of rape in human societies involves much more than merely forcing oneself upon another person; it involves violating another person's rights or failing to obtain adequate consent. This is why consensual sex between an adult and a child is labeled rape, and it also explains why most societies in the past did not consider it conceptually coherent to say that a husband raped his wife. As a result, Dupré argues that we are at risk of grave confusion when we speak of "rape" when talking about the behaviors of animals because they do not have moral concepts in mind when they act.

In addition to highlighting the difficult choices that arise between promoting understanding and minimizing confusion, the term "rape" illustrates one of the deepest ways in which scientific terminology can incorporate values. Because terms like "rape" incorporate not only a descriptive component but also an ethical component, social scientists express moral evaluations when they describe human behavior using these terms. One could try to avoid this problem by abandoning the use of common terms or by trying to redefine them so that they do not have their typical meanings. Nevertheless, Dupré argues that this strategy is typically not helpful in the social sciences. We want these sciences to connect with human needs and concerns, and therefore we want to use terms in ways that are true to their typical meanings. As a result, Dupré claims that many areas of science will be unavoidably value-laden and that this is completely acceptable, given that we want scientists to study the value-laden world of human culture.

Terminology in Toxicology

Even in fields that seem to be completely distinct from human culture, scientific terms often incorporate values. Consider some examples from the field of toxicology, where we might not expect to encounter value-laden terms. Our society currently employs tens of thousands of industrial chemicals, and government agencies and private companies put a great deal of effort into estimating their toxicity. This is difficult to do because it would be unethical to perform experiments in which people were deliberately exposed to significant quantities of potentially hazardous chemicals in a controlled fashion. Therefore, we are forced to perform epidemiological studies (which observe correlations between real-life human exposures and effects) and animal studies. Neither approach provides very exact information, especially about the effects of chemicals at low doses. Government agencies like the Environmental Protection Agency (EPA) and the Occupational Safety and Health Administration (OSHA) currently assume that the toxic effects of harmful chemicals are greater at higher doses than they are at lower doses. They also assume that, at least for chemicals that are not carcinogenic, there is some dose below which the chemicals cease to be harmful.

These assumptions have recently come under fire from a variety of directions. Recall from the earlier chapters of this book that Theo Colborn pioneered the study of endocrine disruption, which occurs when industrial chemicals mimic natural hormones and generate harmful effects as a result. Endocrine-disrupting chemicals appear to challenge the assumptions of regulatory agencies like the EPA and OSHA, insofar as they can generate harmful effects at low doses even though they do not cause problems at higher dose levels. But chapter 1 also pointed out that it is difficult to figure out what to call them. When a US National Academy of Sciences panel issued a report in 1999, it referred to these chemicals as "hormonally active agents." Part of the motivation for using this term was that it seemed less emotionally charged than the term "disruption," and it also eased the impression that hormonally active chemicals are always harmful. But one might worry that speaking only of hormonal "activity" could be misleading in the opposite way. To the extent that endocrine-disrupting chemicals are interfering with the endocrine system, to say that they are merely "hormonally active" might not fully represent the extent of their effects. Speaking of "disruption," "interference," and "activity" returns us to the metaphorical choices that Larson urges scientists to consider more carefully.

Another challenge to previous assumptions in toxicology comes from a phenomenon that is sometimes called "multiple chemical sensitivity" (MCS). Similar to Gulf War Syndrome, those afflicted with MCS appear to become extremely sensitive to a wide variety of chemicals. When they are exposed to a triggering chemical such as a pesticide or a perfume, they suffer a wide range of different symptoms, including headaches, dizziness, indigestion, and difficulty breathing. Because these individuals are so sensitive, and because they experience different symptoms than those normally caused by toxic chemicals, some experts think that MCS is just a psychological problem of mental or emotional origin. As a result, some experts have criticized the term "multiple chemical sensitivity" for giving the impression that the suffering experienced by those with MCS is genuinely caused by chemicals rather than their mental state.

To replace the term "multiple chemical sensitivity," the participants at a 1996 conference in Berlin suggested a new term: "idiopathic environmental intolerance" (IEI). They argued that the term IEI would help advance scientific investigations because it would lessen the impression that chemicals were actually causing the problems. However, a number of experts denounced the new term. They worried that many of the researchers at the Berlin meeting were associated in some way with the chemical industry and thus were purposely trying to develop a term that minimized attention to chemicals as a cause of people's health problems. Their concerns were heightened by the fact that some of the participants at the Berlin conference began to tell the media that the word "idiopathic" meant "self-originated." By using the word "idiopathic" in this way, the term "IEI" became misleading as well because it has not been established that those suffering from this disorder are generating the symptoms themselves.

Further challenges to previous toxicological views come from a phenomenon that is sometimes called "hormesis." In contrast to endocrine disruption, which raised new concerns about the harmfulness of industrial chemicals at low doses, hormesis occurs when chemicals that are normally toxic produce beneficial effects at low doses. For example, there is some evidence that dioxin, the potent carcinogen discussed in chapter 5, might actually decrease the incidence of some tumors when people are exposed to very low levels of it. Edward Calabrese, a professor at the University of Massachusetts, Amherst, has developed a database containing thousands of studies in which he contends that hormesis has occurred. He suggests that the occurrence of hormesis should not be surprising because it makes sense that organisms would find ways to adapt to the harmful substances to which they are commonly exposed. He argues that hormesis is highly "generalizable" across different species and chemicals and that it could have major implications for the way society regulates toxic chemicals. For example, he suggests that regulatory agencies like the EPA should assume that toxic chemicals typically have beneficial effects at low levels, and therefore they need not try so hard to reduce people's exposures. These claims have attracted a great deal of attention from different interest groups, with industry providing funding for his research and public-health-minded scholars challenging Calabrese's findings.

As in the cases of endocrine disruption and MCS, some of the criticisms of hormesis have to do with terminology. Kristina Thayer, a researcher at the US National Institute of Environmental Health Sciences, has argued that it would be better not to even use the term "hormesis" because it is not clear that Calabrese has identified a single, uniform phenomenon that merits a new name. Thayer and her colleagues point out that chemicals have so many different effects on the body that it should not be surprising that toxic chemicals could occasionally have beneficial effects for various reasons. Therefore, they argue that it would be better to say merely that Calabrese has identified "biphasic" or "nonmonotonic" dose-response relationships. These words indicate that the relationship between chemical exposures and bodily responses do not increase uniformly. Unlike the new term "hormesis," however, they do not suggest that a single phenomenon is responsible for all these effects. What is interesting in this case is that the available evidence does not appear to provide compelling reasons to use the term "hormesis" or not, but using the term clearly attracts greater public attention to the idea that normally toxic substances could have beneficial effects. Thus, researchers need to consider whether it is better to use a term that attracts this sort of attention or whether it is better to use a less suggestive term.

Terminology in Medicine

Finally, let us briefly consider how these same terminological issues arise in the field of medicine. In 2015, the US Institute of Medicine (IOM) released a million-dollar report on the classification of chronic fatigue syndrome. According to a 2007 article in the *New York Times*, those suffering from chronic fatigue syndrome have long worried that using this colloquial term for their ailment "has discouraged researchers, drug companies and government agencies from taking it seriously."[11] Thus, some patients have preferred the British term "myalgic encephalomyelitis" because it sounds more medically serious and sophisticated. The 2015 report provided new criteria for defining chronic fatigue syndrome and also proposed a new name: systemic exertion intolerance disease (SEID). The new term "SEID" sounds less medical than the British term, but it is also more descriptive for most people. Moreover, by incorporating the term "systemic," it highlights the fact that even cognitive or emotional exertion can be damaging to patients with SEID.

These decisions about how to refer to chronic fatigue syndrome illustrate the complex range of values that are relevant to choosing medical terminology. Multiple goals are relevant, including not only accuracy, understandability,

11. Tuller 2007.

and comprehensiveness but also the aim of promoting respect and assistance for patients. In some cases, commonly used terms for illnesses clearly generate a lack of respect. For example, chronic fatigue syndrome has sometimes been labeled "yuppie flu," and endometriosis has been called "career woman's disease." It is easy to rule these terms out for value-based reasons, but some cases involve more complex trade-offs. For example, in 2015 the World Health Organization (WHO) called for scientists to avoid naming human infectious diseases in ways that could stigmatize people, places, or animals. For example, using names like Middle East respiratory syndrome (MERS), swine flu, athlete's foot, and Marburg disease would be discouraged under the WHO's new guidelines. But some scientists have argued that much could be lost by replacing these easily understandable names with terms like "filovirus-associated hemorrhagic fever 1" or "BetaCoV-associated SARS 2012." Thus, scientists have to weigh the value of avoiding potential stigma against the value of using catchy terms that are easy to understand and remember.

CATEGORIES AND CLASSIFICATION SCHEMES

In addition to playing a role in scientific framing and terminology, values also have a role to play in choosing scientific categories. As an example, consider racial categories and the ways in which they have contributed to unjustified prejudices and questionable research. In 1994, Richard Herrnstein and Charles Murray published a highly controversial book titled *The Bell Curve*. Their goal was to explain how intelligence varied across American society, to investigate how intelligence contributes to people's success, and to make policy suggestions based on their findings. The core of their argument was that intelligence is an extremely important factor contributing to people's success. Unfortunately, they also concluded that genetic factors are very important in determining people's intelligence and that it is not easy to alter intelligence significantly by changing people's environments. Moreover, while they acknowledged that the evidence is complex, they suggested that these genetic factors play a role in the higher performance displayed by whites and Asians as compared to African Americans on intelligence tests.

Unsurprisingly, their book elicited a host of responses. The renowned scholar of education Howard Gardner claimed that "the science in the book was questionable when it was proposed a century ago, and it has now been completely supplanted by the development of the cognitive sciences and neurosciences."[12] Psychologist Leon Kamin claimed that the book had "disastrous failings," including some "pathetic" data at crucial points.[13] The American

12. Gardner 1995, 61.
13. Kamin 1995, 82.

Psychological Association created a special task force to examine the book's findings. The APA agreed with some findings, such as that intelligence tests tend to predict success in school and in future occupations. But they concluded that there was not convincing evidence for thinking that genetic differences are responsible for the different performances on intelligence tests between whites and blacks.

Renowned paleontologist and science writer Stephen Jay Gould provides valuable historical perspective on *The Bell Curve*. After the book was published, he noted that its interest in providing biological underpinnings for racial disparities in intelligence "is as old as the study of race, and is almost surely fallacious."[14] Gould's classic book *The Mismeasure of Man* summarizes the history of scientific research on racial differences over the past two centuries. He notes that the greatest naturalists of the nineteenth century, including Georges Cuvier, Charles Lyell, and Charles Darwin, all assumed that the African race had lower intelligence than whites. Some scholars of this period thought that the various human races were separate species, while others thought that the "inferior" races ultimately came from the same ancestry as whites but had degenerated significantly. In the first half of the nineteenth century, America's renowned scientist and physician Samuel George Morton collected more than a thousand skulls from different racial groups and concluded that white skulls had larger brain cavities than those of American Indians or Africans. Later anthropologists like Paul Broca amassed a host of further data allegedly indicating that the brains of Africans (and women) were smaller and thus incapable of the same intelligence. Other scientists collected all sorts of bodily measurements (not only of brains but also features like calf muscles, noses, and beards) in order to show that Africans were more physically and intellectually similar to white children than to white adults.

This sordid history of racist research provides a serious cautionary note for those studying intelligence differences in the current era. Gould points out that many of the scientists who performed racist research in the past do not appear to have been purposely trying to "fudge" their data. Nevertheless, he provides numerous examples of subtle but problematic ways in which they interpreted or manipulated the data to support their preexisting biases. As a result, they were able to provide seemingly objective data in support of their racist stereotypes. Ironically, Broca criticized many scientists of his time, accusing them of rejecting the scientific facts because of their egalitarian values. He insisted that one of his most prominent critics "was dominated by a preconceived idea" that all races were equal, and he lamented that "the intervention of political and social considerations has not been less injurious to anthropology than the religious element."[15] While it is tempting to accuse

14. Gould 1995, 5.
15. Quoted in Gould 1981, 116.

Broca of blatant hypocrisy, we already observed in chapter 5 that everyone is prone to this phenomenon of "confirmation bias"—selectively emphasizing or interpreting evidence in order to defend one's initial beliefs. In fact, some of Gould's own research—particularly his criticisms of Morton's measurements of human skulls—appears to have been inaccurate.

Debates about Racial Categories

Given the historical tendency for scientists performing research on racial differences to support their preexisting beliefs with supposedly objective data, it is tempting to argue that scientists should simply abandon racial categories altogether in scientific research. Given our strong social values of promoting equal opportunity and the strong ethical reasons for opposing racism, research that incorporates racial categories seems very unlikely to be helpful. Moreover, the proposal that we abandon racial categories is not just an instance of trying to adopt scientific ideas that we want to be true (i.e., the problem of wishful thinking). Contemporary genetic research indicates that the racial classifications that we emphasize so much in contemporary societies are very tenuous from a biological perspective; comparisons between people of the same racial group commonly find more genetic variation than comparisons between people coming from different racial groups. For example, Mildred Cho, a medical professor and bioethicist, highlights a study that identified around 5% to 10% of white Europeans as having a genetic variant that altered their metabolism of particular pharmaceutical drugs. The study found that only about 1% of Japanese people had the variant. This sounds like a striking racial difference until one realizes that the same variation found between the Europeans and Japanese was also present within the white population. About 10% of whites from northern Spain had the variant, whereas only about 1% to 2% of Swedes had it.

Based on these sorts of findings, Cho argues that the use of racial categories should be abandoned in biomedical research and medical practice. She acknowledges that there are clearly important genetic variations that contribute to diseases, but the racial categories that our societies have created do not track very well with the most significant genetic variations in populations. For example, people in the United States are particularly focused on the racial distinction between "blacks" and "whites," but there is huge genetic variation among African populations that this distinction totally ignores. Moreover, in medical settings people are typically classified into racial groups based on self-reporting or visual inspection by a physician, and these approaches are notoriously inaccurate from a genetic perspective. When one couples the biological imprecision and inaccuracy of racial categories with their potential to promote racist assumptions and prejudices, Cho concludes that it is clearly unwise to employ them.

Nevertheless, other scientists have argued that there can be important benefits to employing racial categories in some cases. For example, medical professor Jay Cohn has argued that even though racial categories are imprecise, they still sometimes correlate with medically significant information. He notes that it does not make sense to look for sickle cell disease in white patients or to look for cystic fibrosis in black ones. He also insists that we should not ignore the evidence that some drugs tend to be more effective in self-identified African Americans than in self-identified whites. And even Mildred Cho acknowledges that racial categories can be helpful in cases where investigators are looking at health disparities across racial groups. Because these disparities are caused in large part by social prejudices against particular racial groups, racial categories are highly relevant to identifying these disparities.

We are left, therefore, with another instance in which values are relevant and necessary for investigators as they decide when, if ever, the benefits of employing racial categories outweigh the harms that they cause. Public health professor S. Vittal Katikireddi and philosophy professor Sean Valles have argued that we should handle these sorts of decisions by thinking about our scientific goals and our ethical goals together. As an example, they appeal to a decision by the US Food and Drug Administration (FDA) to ban blood donations from men who have had sex with men (MSM) since 1977. The motivation was to lower the risk of HIV transmission through blood transfusions. Katikireddi and Valles point out that the use of the MSM category is scientifically questionable because it glosses over the fact that different sorts of male-male sexual behaviors create very different risks of contracting HIV. It is also ethically worrisome because it stigmatizes those in the MSM category. Furthermore, they point out that there are extra benefits to thinking about how these scientific and ethical concerns interact in cases like this. For example, the ethical problem of stigmatizing those in the MSM category exacerbates scientific problems because it inhibits those in this category from communicating openly with healthcare providers and public-health professionals. The scientific problem of using a sloppy category like MSM in turn causes ethical problems because it lessens the opportunities to discuss how different sexual behaviors generate very different risks.

Categories in the Environmental Sciences

These value-laden decisions about how to carve up the world into categories extend beyond the medical and social sciences. Consider, for example, a recent debate in the environmental sciences that focused not on whether to employ particular categories but rather on how to define a socially significant category. We saw in chapter 4 that attitudes toward wetlands changed dramatically during the twentieth century as people started recognizing their value to society. In keeping with the themes of this chapter, it is noteworthy that the term "wetlands" was not commonly used until the 1950s, and it gained prominence partly because scientists, regulators, and environmentalists were looking for a term with more positive connotations than terms like "swamp" or "bog." However, the concept of a wetland has not been easy to define precisely. Wetlands vary in terms of the amount of water present in them and in terms of the soil types and vegetation that they harbor. According to a National Research Council report from the mid-1990s, there was fairly little incentive for scientists to try to adopt a precise classification scheme until they were prompted to do so because of social needs: "Scientists have not agreed on a single commonly used definition of wetland in the past because they have had no scientific motivation to do so. Now, however, they are being asked to help interpret regulatory definitions of wetlands."[16]

Deciding how to define a wetland proved extremely controversial. To see the mixture of values at stake, consider the challenge faced by George H. W. Bush when he became president in 1989. On one hand, he had made a campaign promise that there would be "no net loss" of wetlands under his administration. On the other hand, developers were raising concerns about how this policy hampered their activities. The Bush administration responded to this challenge in an extremely clever fashion. In the 1980s, the four major federal agencies involved in setting policies for wetlands had been working to develop a shared federal definition for this category of land. The Bush administration tweaked their proposed definition in a manner that would have suddenly shrunk the number of lands classified as wetlands—and therefore subject to regulatory protection—by up to one-third. The administration ultimately had to abandon this proposed redefinition because it was met with such intense opposition from the environmental community. Nevertheless, this case illustrates how social values become intertwined with science when scientific categories are important for regulatory purposes. Some critics of the Bush administration argued that he was running roughshod over science, but other commentators pointed out that the scientific evidence does not provide clear guidance about how a category like "wetlands" should be defined. Because the characteristics of these lands vary somewhat continuously in terms of their characteristics, social values may be necessary to decide how to draw the line and decide what should be protected.

16. NRC 1995, 43.

CONCLUSION

We have seen that values are often relevant when deciding how to describe scientific information. The reasons for incorporating values in these decisions correspond to the two justifications given in chapter 1. First, it is often unrealistic to try to find a perfectly value-neutral way of communicating scientific information. The available frames, terms, or categories will subtly promote some values over others; therefore, scientists ought to consider which values are more appropriate to support or emphasize. A second reason is that, when they are deciding how to communicate information, scientists are sometimes faced with multiple goals: ensuring accuracy, preventing misunderstandings, clarifying how their work connects with other areas of science, identifying the social ramifications of their research, and clarifying how their findings relate to people's values and worldviews. Sometimes these goals can stand in tension with each other and with the goal of accuracy, meaning that different frames or terms or categories are better at achieving different goals. In these cases, values are relevant to deciding which goals to prioritize.

Table 6.1 highlights a number of the examples provided in this chapter. They fell into three groups: choices about how to frame scientific findings, decisions about terminology and metaphors, and choices about what categories to employ and how to define them. These examples illustrate the pervasiveness of value-laden decisions when scientists are communicating information. For example, it is striking that in a field like toxicology there have been such pervasive disagreements about terminology. We saw that scientists have debated whether it is better to speak of endocrine disruption or hormonally active agents, whether it is more appropriate to talk about multiple chemical sensitivity or idiopathic environmental intolerance, and whether it is better to speak of hormesis or non-monotonic dose-response relationships.

Of course, we cannot always expect scientists to be aware of the subtle ways that particular frames or terms or categories support some values over others. In most cases, they are probably ignorant about the significance of their language until interest groups highlight its importance. Most scientists are also likely to feel out of their depth as they try to decide what language is most appropriate. As the section on framing discussed, one solution is for scientists to try to "backtrack." In other words, they can take steps to acknowledge when they have made significant choices about their language so that others can recognize the weaknesses of the frames or terms or categories that are used. Another approach is to engage with other scholars and stakeholders to determine how best to communicate scientific information. The next chapter turns to this question of how we can engage a wide variety of stakeholders—citizen groups, policymakers, scientists, and other scholars—to identify value-laden aspects of science and to reflect on how to address them.

Table 6.1 EXAMPLES OF VALUE-LADEN CHOICES IN COMMUNICATING
SCIENTIFIC INFORMATION

Frames	Terms	Categories
In the voles case: genetic determinism, humans are like voles, reductionism, saving relationships, dangers of social manipulation	Metaphors in environmental science: invasive, alien, exotic, native, superabundant, and harmful species	Racial categories: Should they be used and under what conditions?
Discussed by Nisbet and Mooney: social progress, scientific uncertainty, teach-the-controversy, economic competitiveness, religious morality, public accountability, unfair economic burden	Terms related to climate change: greenhouse effect, global warming, climate change, geoengineering, climate intervention, solar radiation management, sunlight reflection methods, albedo hacking	Men who have sex with men: When should the category be employed? Wetlands: How should the category be defined?
Other frames: curing disease, satisfying curiosity, creating technologies	Terms related to biology and social science: monogamy, rape	
	Terms related to toxicology: endocrine disruption, hormonally active agents, multiple chemical sensitivity, idiopathic environmental intolerance, hormesis, non-monotonic dose-response curves	
	Terms related to medicine: chronic fatigue syndrome, yuppie flu, systemic exertion intolerance disease, endometriosis, career woman's syndrome	

SOURCES

Lim et al. (2004) and Walum et al. (2008) are important original sources that present research on voles, humans, and monogamy. McKaughan (2012) and Tucker (2014) provide overviews of this research. Both McKaughan (2012) and McKaughan and Elliott (2013) explore the ways scientific experts and the media have communicated about the research. McKaughan and Elliott (2013) identify the five frames discussed in the chapter and introduce the concept of backtracking. Kavanagh (2007) and Nisbet and Mooney (2007) provide the debate over frames that occurred in *Science*.

Further information about the Chicago Wilderness controversy is available in Gobster (1997) and Shore (1997). Larson (2011) provides his insights about metaphors in the environmental sciences. Elliott (2009, 2011b) explores the significance of terminology in the toxicological cases of endocrine disruption, multiple chemical sensitivity, and hormesis. Gardiner (2004) and Pielke (2007) discuss terminology in the case of climate change, while Elliott (2016a), NRC (2015), and Pierrehumbert (2015) show how values play a role in the terms used for climate geoengineering. McKaughan and Elliott (2013) discuss how the term "monogamy" can be misleading. Dupré (2007) considers the role of values in terms like "rape." Elliott (2009), Grens (2015), and Tuller (2007) examine the terminology used in the case of chronic fatigue syndrome, Capek (2000) discusses endometriosis, and Kupferschmidt (2015) discusses the WHO guidelines for naming diseases.

To read more about the controversy over *The Bell Curve*, see Gould (1981, 1995), Gardner (1995), Herrnstein and Murray (1994), and Kamin (1995). For evidence that some of Gould's criticisms of Morton were inaccurate, see Wade (2011). Discussions over racial categories can be found in Cho (2006), Cohn (2006), and Katikireddi and Valles (2015). Information about the concept of wetlands and efforts at defining them can be found in Elliott (2017). Further work on the role of values in choosing scientific categories and explanatory frameworks can be found in Brigandt (2015), Intemann (2015), and Ludwig (2016).

CHAPTER 7

How Can We Engage with These Values?

O n October 11, 1988, the US Food and Drug Administration (FDA) was shut down by demonstrators from the aggressive activist organization ACT UP (AIDS Coalition to Unleash Power). The crowd of over 1,000 protesters blocked doors, walkways, and a road and chanted "Hey, hey, FDA, how many people have you killed today?" while displaying a black banner that read "Federal Death Administration." The protesters aimed to attract maximum media attention by preparing a press kit and making hundreds of phone calls to the press in advance. Even though the police were apparently instructed to minimize arrests in order to lessen the drama, nearly 180 people were arrested. Some demonstrators tried to enter the FDA building, others displayed T-shirts or posters stating, "We Die—They Do Nothing," someone set off a smoke bomb, others displayed an effigy of President Ronald Reagan, and many people blocked buses designated to transport those who had been arrested.

These demonstrators had several key messages they aimed to communicate to the wider public. One goal was to draw attention to the FDA's slow process for approving new AIDS drugs. While the FDA's methodical drug approval process was designed to protect people from exposure to harmful drugs, patients who were dying from AIDS argued that they should be allowed to take risks with experimental drugs, given that they were likely to die soon anyway. Another message was that placebo-controlled trials were unacceptable in this context. The activists argued that it was ethically inappropriate for people with a life-threatening illness to be given a placebo (i.e., an inactive sugar pill), even if the goal was to advance scientific research. Another important message was that people from all populations—women, men, children, people of color, and the poor—and at all stages of HIV infection should be included in clinical trials, especially given that these trials were sometimes the only source of cutting-edge new therapies.

The activities of ACT UP and other citizen groups ultimately played a significant role in drawing public attention to the challenges faced by AIDS sufferers. What is perhaps most striking about their activism is that they had a significant impact not just on public awareness but also on the practice of biomedical research. Thus, AIDS activism provides a vivid illustration of the ways that community groups can become educated about important scientific issues that affect them and then engage with scientists and government agencies to steer research in directions that address their concerns. In some cases, such as AIDS activism, the relationships between scientists and citizens are largely oppositional. In other cases, scientists collaborate with citizens to design and perform research projects that address community concerns. These efforts, which are sometimes labeled as "community-based participatory research" (CBPR), have become more common in recent decades, as scientists and community groups have come to recognize the benefits of mutual engagement. As discussed in chapter 3, the research collaboration between the Harvard School of Public Health and the citizens of Woburn, Massachusetts, provides an early example of CBPR.

This chapter argues that CBPR and other efforts to promote *engagement* between citizens, policymakers, scientists, and other scholars provide some of the best avenues for addressing the many ways in which values affect scientific research. By "engagement," I mean efforts to interact with other people or institutions in order to exchange views, highlight problems, deliberate, and foster positive change. Four forms of engagement appear to be particularly promising: (1) "bottom-up" engagement between community groups and researchers who can address issues they care about; (2) "top-down" engagement exercises that elicit public input on scientific issues; (3) interdisciplinary engagement between scholars with diverse personal and educational backgrounds; and (4) engagement with the laws, institutions, and policies that structure the practice of scientific research. Each of these four forms of engagement can help to highlight the ways that values are explicitly or implicitly influencing science. They also provide opportunities to challenge or critique value influences that run counter to people's needs and priorities. Finally, they provide opportunities to reflect on the values that would better serve social needs and to incorporate those values in scientific practice.

Before launching into a discussion of all these efforts at engagement, it may be helpful to respond to a potential criticism. A critic of these engagement activities might wonder why people's values are not already adequately represented through the power of the marketplace. As chapter 2 emphasized, two-thirds of scientific research and development in the United States is privately funded. Given that private companies are likely to choose how to invest their research dollars by considering what products people want to buy, one might think that the public's values are already likely to be well represented in most scientific research.

There is certainly some truth to this suggestion, but it also has limitations. First, it applies only to research funded by private companies; we still need strategies for guiding government-funded research. Second, it does little to address the values of those with low incomes. We saw in chapter 2 that many people around the world have very little money to purchase medicine, and therefore pharmaceutical companies provide relatively little funding for diseases that afflict them. Third, products become profitable not only because of consumers' values but also because of government policies that influence whether they can be sold and how much they can be sold for. Thus, we need to consider how to steer these policies. Fourth, some research projects can be profitable even though they do not represent the values of consumers. For example, we saw in chapters 3 and 5 that it can be very profitable for companies to fund misleading research projects that disguise harmful effects caused by products like tobacco, industrial chemicals, or pharmaceuticals. Because of all these considerations, we need to explore multiple ways of incorporating public values into scientific research rather than depending solely on market forces.

AIDS ACTIVISM AND COMMUNITY ENGAGEMENT WITH RESEARCH

The activities of AIDS activists provide excellent illustrations of the first form of engagement discussed in this chapter, namely, bottom-up efforts by citizens to push research in directions they care about. The demonstration at the FDA was only one of many ACT UP activities in the late 1980s and early 1990s. With the motto "Silence = Death," the organization specialized in theatrical efforts to attract public attention. In September 1989, seven members of ACT UP chained themselves to the VIP balcony of the New York Stock Exchange and displayed a banner that read, "SELL WELLCOME." They wanted to challenge the drug company Burroughs Wellcome, which was selling its AIDS drug, AZT, at the exorbitant price of roughly $10,000 for a year's supply. After an article appeared in *Cosmopolitan* magazine in 1988 suggesting that women were typically not at risk from having unprotected sex with male partners infected with HIV, women from ACT UP protested in front of the parent company of the magazine and generated a short documentary and multiple TV appearances. In 1991, during the first Gulf War, several activists entered the studio of the CBS Evening News, jumped in front of the camera, and shouted "Fight AIDS, not Arabs; AIDS is news!" The next day, protesters filled New York's Grand Central Station and displayed banners stating, "Money for AIDS, Not for War," and "One AIDS Death Every 8 Minutes." Other demonstrators delayed the opening of an international AIDS conference and chained themselves to the headquarters of pharmaceutical companies.

The Effects of AIDS Activism on Scientific Research

The combined efforts of ACT UP and other citizen groups ultimately had a significant influence on AIDS research. In his book *Impure Science*, sociologist Steven Epstein describes some of the research efforts that were influenced by the activists. One model, developed in San Francisco, consisted of a County Community Consortium (CCC) that helped to organize community-based drug trials. As part of their work with patients, primary-care doctors in the CCC provided drugs and collected data about their efficacy. While the CCC helped to integrate research with the primary care that patients were receiving, it did relatively little to incorporate patients in decision-making. Another model, developed in New York City, broke down this barrier. The NYC Community Research Initiative (CRI) allowed AIDS patients to collaborate with activist physicians in making decisions about what trials should be conducted and even how they should be designed. Some of the crucial issues at stake included whether placebos should be used and what criteria would be used for including or excluding potential research subjects. Pharmaceutical companies became sufficiently impressed with the sophistication of the CRI that they began to collaborate on drug studies, and in 1989 the FDA approved an AIDS-related treatment based solely on community-based research from the CCC and CRI.

In addition to developing their own community-based research, activists even worked to change research and policy practices at the US Food and Drug Administration (FDA) and the National Institutes of Health (NIH). ACT UP initially had a very tense relationship with these government agencies, with activists accusing President Ronald Reagan and prominent government health officials of "genocide-by-inaction" based on the criticism that they were not taking adequate steps to address the AIDS epidemic. Prominent activist Larry Kramer wrote that these figures were "equal to Hitler and his Nazi doctors performing their murderous experiments in the camps— not because of similar intentions, but because of similar results."[1] One of the activists' greatest complaints was that the FDA was preventing dying patients from accessing new AIDS drugs because it demanded so much evidence before approving them.

1. Quoted in Epstein 1996, 221.

The activists were also critical of the NIH, and specifically the National Institute of Allergy and Infectious Diseases (NIAID), because of the ways it designed studies on new AIDS drugs. As mentioned in the introduction to this chapter, one of the activists' primary concerns was that the standard method of testing new drugs was to assign half the experimental subjects a placebo, which meant that they were denied the potential benefits of the new drugs. The activists also complained that the trials were designed in such a way that many people were not allowed to participate—for example, those who had been taking other drugs or who had additional illnesses. Sometimes the studies even excluded women. Biomedical researchers justified these exclusionary practices, as well as the use of the placebos, because it enabled them to obtain more straightforward, "cleaner" study results that were easier to interpret. The activists pointed out that if the only way to access experimental medications was to participate in a drug study, then these exclusionary practices raised significant ethical concerns. They also argued that the results of the studies might not be relevant to real-life patients, who did not have the characteristics of the "pristine" study subjects. Even worse, the activists pointed out that if study subjects thought they might be in the placebo group of a study, they were likely to cheat and obtain experimental drugs through underground "buyers clubs," thereby invalidating the results of the studies.

In an effort to change these research practices, many activists went beyond staging demonstrations and began educating themselves about the details of AIDS research. Their ability to bring themselves up to speed was impressive. As the director of a New York drug buyers club noted: "When we first started out, there were maybe three physicians in the metropolitan New York area who would even give us a simple nod of the head. Now, every day, the phone rings ten times, and there's a physician on the other end wanting advice. [From] me! I'm trained as an opera singer!"[2] Louis Lasagna, an expert on clinical trial methodology who chaired a government committee on AIDS drugs, emphasized the contrast between the demeanor of the AIDS activists and their scientific knowledge. He recalled that they "came dressed in any old way almost proud of looking bizarre." Nevertheless, he said, "I'd swear that the ACT UP group from New York must have read everything I ever wrote."[3] Epstein argues that by learning the language of the biomedical researchers "activists increasingly discovered that researchers felt compelled, by their own norms of discourse and behavior, to consider activist arguments on their merits."[4]

2. Quoted in Epstein 1996, 229.
3. Quoted in Epstein 1996, 232.
4. Epstein 1996, 231-232.

Based in part on their growing scientific sophistication, the activists made extraordinary strides in altering the practice of government agencies. Anthony Fauci, the head of the NIAID (who had previously been called an "incompetent idiot" by Larry Kramer),[5] began to accept many of the activists' suggestions. He noted that the activists had "an extraordinary instinct ... about what would work in the community ... probably a better feel for what a workable trial was than the investigators."[6] The NIAID began to employ study designs that did not require placebos, which lessened the incentive for research subjects to "cheat" and take illicit drugs while participating in studies. The activists also encouraged researchers to ease some of their strict requirements for excluding patients from clinical trials, which in turn made it easier to enroll research participants. The NIAID also began to allow activists to serve on the committees that determined which treatments and studies to pursue. Even the FDA softened their strict stance on drug approvals and adopted a "conditional approval" process, which allowed some drugs to be approved with less evidence than normal, as long as the manufacturers took subsequent steps to assess their effectiveness.

It is striking to see how many of the different roles for values that we have explored in this book were addressed by the work of these AIDS activists. First, they agitated for greater investments in research to address AIDS, thereby influencing societal decisions about which research topics to prioritize (see chapter 2). Second, part of the reason that activists pushed to be on the NIAID committees that made decisions about AIDS treatments and studies was that they felt the NIAID had put too much focus on investigating antiviral drugs and not enough emphasis on drugs that could help minimize infections. Thus, they were raising concerns about the specific questions that were being asked about the topic under investigation, as discussed in chapter 3. Third, when the AIDS activists pushed to ease the exclusion criteria for clinical trials so that more people could participate, they were influencing decisions about research aims, as discussed in chapter 4. Specifically, they were arguing that it was more important to design studies that represented the messiness of the real world rather than trying to obtain the very "clean," easy-to-interpret data that biomedical researchers preferred. Fourth, the activists' clashes with the FDA over the approval of new drugs was a debate over the appropriate levels of evidence to demand, which we examined in chapter 5. Fifth, many activists challenged early definitions of AIDS because those definitions focused on the ways the disease presented itself in men and ignored the unique features of AIDS in women. These are precisely the sorts of terminological issues discussed in chapter 6.

5. Quoted in Epstein 1996, 236.
6. Quoted in Epstein 1996, 249.

AIDS activism is just one of many cases in which community groups have engaged with scientists to explore the value-laden aspects of scientific research. Sometimes these engagements have been largely oppositional, such as when AIDS activists challenged early research funded by the NIH. In other cases the engagements have been more collaborative. As far back as the nineteenth century, progressive reformers like Jane Addams and Alice Hamilton worked with community members at Chicago's Hull House to identify pollution problems and workplace hazards. In the 1970s and 1980s, the cases of Woburn, Massachusetts, and Love Canal, New York, drew new attention to the benefits of community research partnerships. At Love Canal, concerned citizens collected information about disease in their community and ultimately uncovered evidence of toxic waste pollution that forced the government to relocate them.

At present, organizations like the Louisiana Bucket Brigade work with citizens to help them collect evidence of air pollution in their neighborhoods. A number of other citizen groups and nongovernmental organizations (NGOs) around the world are working with small-scale farmers to perform research on agricultural techniques that meet their local needs. Sometimes, community groups have even collaborated with industry, such as when a community advisory council in Alaska worked with the petroleum industry after the 1989 *Exxon Valdez* oil spill to study safer methods for moving ships through Prince William Sound. There are now also major efforts by funding agencies like the NIH to foster partnerships between community groups and academic researchers.

While collaborative relationships between scientists and citizen groups can sometimes be more fruitful than oppositional ones, they can also become problematic when citizens lose too much power over the direction of the collaboration. Sometimes the values of citizens and the values of scientists diverge, such as when scientists want to pursue investigations that are more interesting from an academic perspective while citizens want to focus on more applied questions that can help them address injustices in their communities. One response to this challenge is for community groups to take control over research collaborations rather than letting university scientists guide them. For example, the West End Revitalization Association (WERA), a community-based organization in North Carolina that was founded in the 1990s to address public-health and environmental problems in low-income African American communities, found that it was most effective to pursue research grants on its own and to hire university faculty as partners and consultants. By doing so, WERA was able to determine which questions to ask, maintain ownership of the collected data, protect the interests of the local community, and maintain people's trust. In some cases, this "community-owned and managed research"

(COMR) model may be a helpful compromise between CBPR (which is typically guided by academics) and more oppositional relationships between communities and scientists.

Sociologist Abby Kinchy has described another form of compromise between oppositional and collaborative relationships. This form of engagement, which she calls an epistemic boomerang, occurs when activists influence scientists so that they represent the activists' concerns to indifferent or oppressive governments. This approach can be particularly helpful for citizen groups because contemporary public policy is often based heavily on scientific analyses, which can leave ordinary citizens at a disadvantage when they try to alter policy decisions. Kinchy illustrates the notion of an epistemic boomerang by describing a case in which Mexican activists who opposed genetically modified (GM) corn convinced an expert advisory committee to express many of their concerns. The committee was formed in response to persistent citizen complaints about the spread of GM corn into Mexican agriculture, and it was scheduled to present its preliminary findings at a public symposium in March 2004. As Kinchy puts it, "NGOs and rural groups used the public scientific meeting as an opportunity to demonstrate their opposition to GE maize [i.e., corn] through testimony, protest signs, theatrical interventions, and the placement of colorful mosaics of maize on the floor, all of which disrupted the planned sequence of events for the symposium."[7] According to Kinchy, these demonstrations had a powerful impact on the experts who attended, and they ultimately expressed many of the citizens' concerns in their final report.

One of the striking features of these cases in which communities engage with scientists is that they can contribute not only their values but also their knowledge. Given what we have learned in this book about the tight connections between facts and values, it should not be surprising that the values and the knowledge of communities are closely related. Because community members have unique experiences, their values sometimes differ from those of scientific experts, and those values lead them to collect different sorts of information or to ask different questions. Some scholars now use the term "local knowledge" to refer to the unique information and insights that citizens can contribute to scientific research. For example, agricultural workers sometimes know more details about how pesticides are actually used than academic scientists, and thus their insights can be valuable when assessing pesticide risks. Indigenous communities often have unique forms of "traditional ecological knowledge" (TEK) that they have accumulated over centuries of living in a particular location. And as we saw in the case of the Madison Environmental Justice Organization (chapter 3), local community groups often have privileged information about the foods people eat, the places they

7. Kinchy 2010, 188.

go, and the hazards to which they are exposed. Thus, efforts to engage citizens in research are sometimes important not only as a way to incorporate their values in the research enterprise but also as a way of benefiting from their unique sources of knowledge.

FORMAL ENGAGEMENT EFFORTS

A second approach to creating engagement is to initiate it from the top down through organized, formal efforts. A good example of this approach comes from a burgeoning new field called nanotechnology, which focuses on the manipulation of matter at the molecular level. In 2002, bestselling author Michael Crichton wrote a novel called *Prey*, which explored the potential hazards associated with this area of research. Perhaps best known for writing the book that inspired the blockbuster movie *Jurassic Park*, Crichton wrote more than twenty novels that touched on various facets of science and technology. In *Prey*, he investigated the idea that scientists could potentially create "assemblers" consisting of tiny machines that manipulate matter at the molecular level. In the novel, Crichton portrays swarms of solar-powered, self-sufficient assemblers that escape from a laboratory and reproduce. These assemblers develop the ability to engage in coordinated, intelligent activities. They infiltrate human bodies and alter their behavior, and groups of assemblers are even able to swarm together to mimic the appearance of other humans.

While many aspects of Crichton's story are highly unrealistic, some commentators have proposed that scientists could indeed develop molecular-scale machines that pose important social and ethical issues. Nanotechnology involves much more than the dream of creating tiny machines, however. The term "nanotechnology" stems from the word "nanometer," which refers to a billionth of a meter, or roughly the length of 10 hydrogen atoms. Most current examples of nanotechnology involve efforts to create tiny particles (100 nanometers or less) made out of ordinary materials like gold or silver or carbon. These particles often have unique properties that can be used for developing exceptionally strong but light-weight materials, self-cleaning surfaces, better solar cells, advanced computer chips, and other new applications. In medical contexts, nanotechnology could also be used for developing tiny sensors or for encapsulating drugs in special molecular structures that regulate their release in particular tissues.

At the end of his presidency, Bill Clinton launched a National Nanotechnology Initiative (NNI) in the United States, which coincided with significant worldwide investments in nanotechnology around the world. Between 2001 and 2015, the NNI invested more than $20 billion in nanotechnology research activities across 20 different government departments and agencies. As part of this funding, the NNI included money to investigate environmental health and safety (EHS) issues as well as ethical, legal, and societal issues (ELSI). For example, a number of concerns have been raised about the possibility that particles in the nano-size range could have different toxicity properties than larger particles and that these toxic properties could be difficult to predict. Others worry that nanotechnology could be used to construct tiny sensors that threaten people's privacy or to create new forms of human enhancement or military technology that raise ethical concerns. Finally, some commentators do indeed raise the concern that nanoscale machines could run amok and start uncontrollably manipulating the earth at the molecular level, ultimately turning everything into a "gray goo."

Unfortunately, it is somewhat more difficult to promote public engagement around a scientific topic like nanotechnology as compared to community health problems like leukemia, asthma, or AIDS. When community health threats are present, there are typically concerned citizens who are motivated to engage with the research community. The situation is somewhat more complicated when researchers are studying new technologies or topics that do not immediately affect local communities. In these cases, citizens often know relatively little about the science, and they are not particularly motivated to work with scientists. Moreover, the few citizen groups that are highly motivated to engage with the scientific community might end up holding extreme views—such as intense enthusiasm or distrust of new technologies—that do not represent the full range of public values.

To address this challenge, social scientists and policymakers have explored a number of techniques for eliciting public input about important value judgments in scientific research. In the case of nanotechnology, for example, some scholars have performed survey research to assess public enthusiasm about nanotechnology and to determine which risks or dangers seem particularly significant. Unfortunately, a disadvantage of this approach is that most people are very unfamiliar with nanotechnology, and thus they have little basis for providing informed opinions. Therefore, techniques that allow citizens to engage with experts and then discuss important issues as a group tend to generate much more thoughtful perspectives. From 2006 to 2009, the European Commission provided funding for a project called DEEPEN (Deepening Ethical Engagement and Participation with Emerging Nanotechnologies), which employed more intensive efforts at citizen engagement. One of the more interesting approaches involved gathering input from ten citizen groups (six in the United Kingdom and four in Portugal), with about seven individuals in each group. One group came from a church, another group was interested in organic products, another consisted of supporters of new technologies, another involved workplace leaders, and still others were focused on environmental issues or consumer rights.

Each group met on two different occasions to discuss technologies in general and to react to reading material about nanotechnology. Then each group was paired with another group for a Saturday workshop. During the morning of the workshop, each group worked individually to discuss their most significant issue or concern related to nanotechnology, and they developed a presentation or performance about the issue. In the afternoon, the paired groups presented their performances to each other and discussed their perspectives on the future of nanotechnology. According to the organizers, they purposely designed an approach that gave people the opportunity to create "theatrical" performances for each other because this can "harness unexamined, affective and intuitive ethical responses."[8] By synthesizing information from these workshops and other events at which citizens shared their perspectives, the organizers identified five narratives that people consistently expressed: (1) be careful what you wish for; (2) opening Pandora's box; (3) messing with nature; (4) kept in the dark; and (5) the rich get richer and the poor get poorer.

In the United States, the National Science Foundation funded two Centers for Nanotechnology in Society (CNS) in 2005. One goal of these centers was to gather public perspectives on the future course of research in this area. The CNS at Arizona State University created a National Citizens' Technology Forum (NCTF) to gather public perspectives on nanotechnology, focusing especially on how it could potentially contribute to human enhancement (e.g., improved sight or hearing or strength). The NCTF was inspired by an approach to public engagement called a consensus conference, which was originally developed in Denmark and then practiced in a number of other countries. In a consensus conference, a group of citizens is selected to provide its perspectives on an important area of science or technology. They read background material about the topic and meet several times to learn more about the relevant issues. Then they meet for a weekend conference and have the opportunity to interview experts, after which they develop a report summarizing their perspective on the topic and the major issues that they think policymakers ought to address.

Previous experiments with consensus conferences have shown that citizens can produce very insightful and well-informed documents through this process. However, this approach is easier to justify in a small, relatively homogeneous country like Denmark as compared to a huge, very heterogeneous country like the United States. The NCTF employed a modified format that incorporated 74 people from six different locations across the United States. Over the course of a month, the participants read background materials, met face-to-face with other participants at their local site, and deliberated with participants and experts from distant locations through an online system. During a final weekend, the participants met again in a face-to-face format at each of the six sites to create reports on their conclusions. According to David Guston, the director of ASU's CNS, the NCTF yielded thoughtful reports that

8. Davies et al. 2009, 22.

revealed support for using nanotechnology to improve medical therapy but discomfort with the idea of using it to pursue human enhancement.

Scholars continue to explore ways to improve these engagement efforts. At ASU, they are now experimenting with what they call "material deliberation."[9] Rather than seating a group of citizens around a table to discuss science and technology, they are trying to elicit people's perspectives through hands-on experiences, such as playing games, going on walks to discuss urban infrastructure, engaging with artwork, or participating in simulations. Science museums are also collaborating to develop creative, interactive exhibits that inform people about advances in fields like nanotechnology while introducing major social and ethical issues that accompany the new scientific developments. Some of these efforts are focused less on obtaining immediate feedback from citizens and more on building the public's capacity to influence the future development of science and technology. Another approach, which builds on the Danish consensus conference model, is the World Wide Views engagement effort. It brings groups of local citizens together on the same day in countries around the world to discuss common questions about pressing social issues. In 2009, 4,000 citizens from 38 countries answered questions about their views on global warming, and in 2012, 3,000 citizens from 25 countries deliberated about biodiversity loss.

Of course, while these efforts to engage members of the public in informed deliberation about science and technology can yield valuable insights, they also fall prey to a variety of weaknesses. For example, they seem to be much better at addressing values in some aspects of science, such as the choice of research topics (chapter 2), than in other aspects of science, such as making decisions about how to handle uncertainty or describe information (chapters 5 and 6). In fact, a major worry about these engagement efforts is that citizen deliberations could be swayed in advance by the way the background material about a topic is initially framed and communicated. This illustrates a general concern about these efforts, namely, that there is typically a significant asymmetry of power in the ways they are designed. Government employees or academic researchers decide how to structure them, and citizen responses are influenced by that structure.

Another worry about formal engagement efforts is that it is somewhat unclear exactly how they should influence the normal policymaking process. Because only a small number of people are involved in them, it would be unwise to give these exercises too much weight in deciding what policies to implement, but they should clearly play some role in policymaking. Related to this point, it is often unclear exactly what these efforts are trying to accomplish. One possible goal is to influence policy directly. A somewhat different

9. Guston 2014, 56.

goal is to educate citizens about important issues. One might also try to use these exercises to increase the public's capacity for civic involvement. Another goal is to influence scientific practice. Finally, these engagement efforts might be designed to increase public acceptance of new areas of science and technology so that later conflict is minimized. Going forward, it is probably best to recognize that public engagement can serve multiple purposes, and different approaches to engagement are probably best at achieving different goals in different contexts.

DIVERSE, INTERDISCIPLINARY SCHOLARSHIP

A third approach is to promote engagement between interdisciplinary groups of scholars with diverse backgrounds. Consider the example of Kristin Shrader-Frechette, an endowed professor of philosophy and biological sciences at the University of Notre Dame who has authored hundreds of articles on the role of ethics and values in science and technology. What is perhaps most striking about Shrader-Frechette's work is that much of it has involved pro bono efforts on behalf of local communities struggling with environmental pollution. As the Director of Notre Dame's Center for Environmental Justice and Children's Health, she works with students in her courses to provide evidence of links between community health problems and toxic-waste dumps or other sources of pollution.

Her book *Taking Action, Saving Lives* opens with an example of the sorts of cases she often studies. Emily Pearson, from the town of Hammond along the northern border between Indiana and Illinois, was diagnosed with brain cancer when she was three years old, and she died in 1998 at the age of seven. Much like the other cases of citizen action discussed in this book, her mother Gwen worked with other parents to identify more than 100 nearby children diagnosed with cancer, and she founded Illiana Residents Against Toxico-Carcinogenic Emissions (IRATE). They believed that much of the cancer incidence could be linked to emissions of ethylene dichloride (EDC) and vinyl chloride (VC) from the nearby Ferro Chemical Plant. Because of cancer concerns, the US government had limited the plant to annual emissions of 50,000 pounds of compounds like EDC and VC, but these limits were not enforced, and in the 1990s the plant was releasing almost 2 million pounds of EDC per year.

Despite these concerns, a toxicologist from Ferro argued on the basis of data from a local air-monitoring facility that the company's emissions were not related to local childhood cancers. The US Agency for Toxic Substances and Disease Registry (ATSDR) and the Indiana Department of Public Health agreed with Ferro. In her book, Shrader-Frechette argues that these analyses were based on numerous questionable assumptions and that these agencies should not have given Ferro a "clean bill of health." For example, she points out that the air-monitoring facility was not located downwind of Ferro, it did not measure exposures in key residential areas, and it did not address potentially high short-term exposures to volatile chemicals like EDC. She also criticizes the analysis by the Department of Public Health because it focused on cancers in the entire county and not just those near the plant, it examined only cancers in Indiana and not in neighboring Illinois, and it averaged together all cancers instead of focusing on the rare cancers of those living near the plant.

Highlighting Value Judgments

Shrader-Frechette's work exemplifies how engagement between interdisciplinary scholars with diverse backgrounds can help to address important value judgments. The worries she highlighted about the Ferro Chemical Plant are precisely the sorts of issues that we discussed in chapter 3, where we saw that scientists need to make decisions about what methods to use, what assumptions to employ, and what specific questions to ask. Because Shrader-Frechette is a philosopher with training not only in ethics but also in math and science, she is particularly well equipped to uncover ways in which values influence scientific work on environmental pollution.

Admittedly, someone could argue in the case of the pollution coming from the Ferro Plant that one does not need to consider values in order to criticize the analysis by the ATSDR and the Indiana Department of Public Health. From a strictly value-neutral perspective, one could argue that their data were so limited that they should have declared the evidence to be ambiguous rather than claiming that there was enough evidence to exonerate the plant. As we saw in chapter 5, however, deciding how much evidence to demand before exonerating the plant (as opposed to declaring the evidence to be ambiguous) is still a value-laden decision. Thus, it is extremely valuable for scholars like Shrader-Frechette to point out how different assumptions, methodologies, and standards of evidence—which in turn support some social values over others—can result in different conclusions.

Shrader-Frechette has performed this sort of analysis in a variety of cases involving chemical pollution, conservation biology, and nuclear waste disposal. For example, she examined in detail the US Department of Energy (DOE) assessments of Yucca Mountain, Nevada, as a site for building the country's first permanent, high-level-nuclear-waste repository. Between 1978 and 2011, more than $10 billion went toward studying Yucca Mountain. The stakes surrounding the site were huge because the federal Nuclear Waste Policy Act of 1982 tasked the DOE with building a permanent facility for storing spent waste from nuclear power plants, but it has been extremely difficult to identify a location where citizens feel comfortable storing waste that will be lethal for tens of thousands of years into the future. Many government experts regarded the Yucca Mountain site as ideal because it received little rainfall, it did not experience much volcanic or seismic activity, it was not close to large population centers, and it was already owned by the federal government for nuclear-weapons testing. Even though the DOE approved the site in 1992 and began excavations, the state of Nevada and a variety of citizen groups created a string of legal challenges throughout the following two decades, and in 2011 President Obama ultimately halted funding for Yucca Mountain.

Much as she did in the case of the Ferro Chemical Plant in northwest Indiana, Shrader-Frechette argued that the DOE analyses of Yucca Mountain incorporated a variety of questionable assumptions that merited additional scrutiny. For example, she claimed that the DOE studies did not adequately account for numerous uncertainties arising from the extremely long timescale (namely, hundreds of thousands of years) envisioned for the site. Some of the most important uncertainties involved potential future seismic or volcanic activity, the possibility that there could be precious materials at the site that would attract future human excavations, the failure rates of waste casks, and the possibility that fractures in the rock could allow groundwater to enter the facility and carry out radioactive material. Moreover, when the DOE employed models to try to address some of these uncertainties, she contends that they did not adequately simulate the actual conditions at Yucca Mountain. By performing these analyses, scholars like Shrader-Frechette create opportunities to scrutinize the many assumptions involved in research projects that inform public policy. For example, even if one were to conclude that the DOE's models were adequate, Shrader-Frechette's analysis would help to clarify the crucial assumptions that one would be accepting as part of the models.

Other Models for Interdisciplinary Engagement

Like the citizen groups described earlier in this chapter, Shrader-Frechette has incorporated not only oppositional but also collaborative approaches into her work. For example, she has served with scientists on numerous government committees associated with the Environmental Protection Agency, the National Academy of Sciences, and the International Commission on Radiological Protection. Another philosopher, Nancy Tuana, has also modeled important ways that scholars from the humanities and social sciences can collaborate with natural scientists. As Director of the Rock Ethics Institute at Penn State University, Tuana has worked with climate scientists who develop Integrated Assessment Models (IAMs) of climate change. As discussed in chapter 4, her goal has been to help them recognize ways in which their modeling choices end up supporting some values rather than others. She calls this "intrinsic" or "embedded" ethics.

Recall from chapter 4 that IAMs bring together information from the physical sciences and the economic sciences in an effort to predict the overall consequences of policy decisions about spending money to mitigate climate change now versus waiting to adapt to climate change in the future. We saw that these IAMs incorporate significant choices that philosophers like Tuana are well-equipped to point out. For example, an IAM that focuses only on the overall future costs and benefits of a particular policy on climate mitigation may fail to identify very significant inequities in the benefits and burdens borne by the poorest members of society as opposed to the wealthiest. We also saw that the construction of IAMs require decisions about how to compare future costs and benefits relative to present costs and benefits, and they require placing monetary values on natural catastrophes such as floods and droughts. Some IAMs also incorporate requirements that particularly dire events—such as the collapse of the current in the North Atlantic Ocean that maintains Western Europe's relatively temperate climate—must not be allowed to exceed a particular probability. Tuana has worked with climate scientists and their graduate students to highlight the significance of these modeling choices.

Some academics have created formal structures to help facilitate interdisciplinary collaborations between natural scientists and other scholars who study the role of values in scientific research. For example, science-policy scholar Erik Fisher developed an NSF-funded project that "embedded" social scientists and humanists in research labs with scientists and engineers. One of the primary goals of the embedded scholars was to ask constructive questions about precisely the sorts of issues explored throughout this book: why are particular modeling choices being made, why is the research being directed toward achieving the aims that it is, why are specific research questions being asked (as opposed to others), and why is the research enterprise being framed in the way that it is? Fisher developed the term "midstream modulation" for the process of reflecting on these questions and adjusting research efforts accordingly. He has argued that midstream modulation can be helpful both for promoting social values and for helping scientists to accomplish their narrower scientific goals of producing high-quality, publishable work.

Philosophers Michael O'Rourke and Stephen Crowley have pioneered another strategy for promoting thoughtful, interdisciplinary reflection within research projects. Their approach focuses on interdisciplinary groups of scientists who are already collaborating together (or who are preparing to collaborate). They created an instrument called the "Toolbox," which consists of a number of philosophical statements about scientific research, such as "research interpretations must address uncertainty" and "determining what constitutes acceptable validation of research data is a value issue."[10] To utilize

10. O'Rourke and Crowley 2013, 1952.

the Toolbox instrument, each of the collaborators in a research group record the extent to which they agree with the statements, and then they spend time together in a dialogue about their responses. After the dialogue, they fill out the Toolbox instrument a second time and record their reactions to the experience. O'Rourke and Crowley have found that this process can be very valuable when scientists from different disciplines are working together. They often find that important words like "replication," "representation," or "model" have different meanings in different fields. Moreover, scientists working in these different fields can have very different expectations about how to confirm their findings or whether values have a legitimate role to play in particular aspects of scientific practice. Thus, the Toolbox instrument can facilitate critical reflection about many of the issues discussed throughout the preceding chapters of this book.

There are a number of other intriguing ways that engagement between scholars with diverse backgrounds can help to promote more thoughtful reflection about the ways that values influence science. While this section has focused especially on promoting disciplinary diversity (e.g., bringing social scientists and humanists together with natural scientists), the history of science indicates that increasing the gender and racial diversity of the scientific community can be extremely valuable as well. In fields like anthropology and biology, the inclusion of female scientists has often helped to uncover questionable ways in which values influenced scientific methods, assumptions, and theories. Chapter 4 discussed how a group of female anthropologists challenged "man-the-hunter" theories of human evolution. Similarly, some female (and male!) biologists have highlighted weaknesses in research projects designed to identify biological reasons for stereotypical male qualities such as promiscuousness, competitiveness, and aptitude for mathematics. Women in fields like primatology have also pointed out observational methods that tended to neglect important activities of female animals.

Thus, promoting diverse, interdisciplinary research efforts appears to be a powerful way to uncover hidden value judgments and promote fruitful reflection on them. We have seen that one way to achieve this goal is for scholars with diverse viewpoints to act as critics "from afar," questioning the assumptions and methodologies underlying other research projects. Another approach is to bring a diverse group of participants together to foster collaborations. One of the important remaining questions, though, is how to create institutional settings that promote this sort of interdisciplinarity and diversity. Some steps have already been taken. Funding agencies like the NSF and NIH are encouraging interdisciplinary grant applications. Universities are encouraging collaborations across disciplines and even hiring scholars jointly across departments—partly in response to the pressure from funding agencies to promote interdisciplinarity. Efforts are being made to identify and address the factors that prevent more women and underrepresented minorities from persisting in scientific fields. The next section considers these and other institutional factors that can help in addressing the role of values in scientific practice.

INSTITUTIONAL ENGAGEMENT

A fourth approach focuses on engagement between different groups of people and the laws, institutions, and policies that influence scientific research. This form of engagement is somewhat different from the first three because it focuses not just on engagement between different groups of people but on engagement between people and the laws or policies that affect them. Thus, this fourth form of engagement overlaps with the first three, insofar as all three can be directed not only at changing individual research projects but also at changing institutions or laws that affect many different research projects. For example, patent policies can have a huge influence on the course of scientific research. As we saw in chapters 2 and 3, the ability to patent an invention is a major incentive for performing research to develop it because it provides the inventors with the opportunity to reap significant profits. We also saw that patent policies can result in expensive drugs that are not particularly novel if the policies are not designed strategically. Thus, policies that determine which innovations are patentable can support some values rather than others, insofar as they influence which research topics and questions are pursued. Efforts to alter these policies—through bottom-up community engagement, top-down formal engagement, interdisciplinary efforts among scholars, or any other approach—provide examples of this chapter's fourth form of engagement.

Consider the story of a mouse genetically engineered in the early 1980s by researchers at Harvard University so that it would be particularly susceptible to developing cancer. The Oncomouse, as it came to be called, was ideal for studying cancer biology as well as potential cancer treatments. Nevertheless, this discovery was soon engulfed in controversy. The Harvard researchers received a US patent on their new mouse and licensed it to DuPont, but this patent became a touchstone for international debates. On one side, many members of the scientific community felt that DuPont charged overly high prices for access to the Oncomouse and thereby inhibited the progress of scientific research. On the other side, many citizen groups felt that it was ethically unacceptable for a living thing—especially a familiar animal like a mouse—to be patented at all. These debates over the Oncomouse illustrate the importance of engaging with the institutional structures and rules that can play a significant role in determining which values influence scientific practice.

The patent for the Oncomouse was submitted only a few years after a landmark 1980 Supreme Court decision, *Diamond v. Chakrabarty*, which opened the door for patents on living organisms. This 5-4 decision was itself controversial, but it involved the patenting of a genetically modified bacterium; the notion that a mammal could be patented was even more contentious. Jeremy Rifkin, a prominent environmentalist and anti-biotechnology activist, led a coalition of groups that argued against the patenting of animals. They employed a number of arguments, including appeals to the suffering associated with genetically altered animals, inherent ethical concerns about being able to "own" living things, and worries about the loss of biodiversity. Nevertheless, the US Patent and Trademark Office (PTO) insisted that US patent law does not include ethical and social concerns as relevant to patenting decisions, and it ultimately granted a patent on the Oncomouse in 1988. Rifkin was so frustrated by the PTO's approach that he filed a patent of his own for an embryo composed of a mixture of human and non-human cells. His application was a public-relations stunt intended to draw public attention to what he regarded as an absurd failure by the US patent system to consider ethical and social considerations when evaluating patent applications.

Science-policy scholar Shobita Parthasarathy has argued that we can learn a great deal about how institutional patent policies block or allow value influences by performing comparisons across different countries. She has argued that while US policies blocked the values of citizens from influencing patent decisions over the past 30 years, the system in Europe has allowed greater citizen input. One of the biggest reasons for this difference is that the European patent law includes a clause stating that patents can be blocked if they violate morality or public policy (translated into French as "ordre public"). Thus, European citizens can appeal to moral considerations (such as the inappropriateness of patenting living things) as a basis for rejecting patent applications.

When the Oncomouse patent went to the European Patent Office (EPO) and citizen groups challenged it, they were able to exert some influence because of the "ordre public" clause. The EPO granted the patent in 1992, but they were forced to justify their decision partly on ethical reasoning, arguing that the overall consequences of allowing the patent were better than the consequences of rejecting it. Opposition figures were subsequently able to force the EPO to respond to their social and ethical objections. While they did not manage to eliminate the patent, they were ultimately able to restrict its scope after a decade of legal debates, and they successfully rejected other patent applications based on the ordre public clause. In contrast, citizens in the United States cannot explicitly appeal to concerns about values or ethics when attempting to influence patent decisions, but the PTO's patent decisions still implicitly end up supporting some values in research while suppressing others.

Other Approaches for Structural and Institutional Engagement

Rifkin's efforts to engage with the patent system provide a particularly vivid illustration of the ways people can engage with scientific institutions, laws, or policies in an effort to alter the values that influence them, but there are a number of other important examples. Chapter 2 briefly described Thomas Pogge's proposal for a Health Impact Fund that would provide a financial incentive for pharmaceutical companies to do research that serves social values (namely, helping the poor). A number of other mechanisms can also be used to alter incentives for private companies. For example, governments and nongovernmental organizations (NGOs) can pool money in order to provide advance market commitments (AMCs) to pharmaceutical companies. In an AMC, the donors promise to purchase a large quantity of a drug or medical treatment at a specified price in order to make it available to those with low incomes. This can provide enough financial incentive for companies to develop treatments that would not otherwise be profitable. In 2009, for example, a number of organizations, including the World Health Organization, the World Bank, the Bill and Melinda Gates Foundation, and several national governments, launched an AMC designed to encourage the development of vaccines for diseases that primarily afflict the poor.

Government regulations can also steer scientific research toward serving values that would otherwise be neglected. For example, as we saw at the end of chapter 3, philosopher Carl Cranor has argued that the US Toxic Substances Control Act (TSCA) previously encouraged chemical companies to avoid collecting information about the potential hazards associated with their products. This is because TSCA allowed companies to market industrial chemicals without showing that they were safe. In contrast, insofar as the 2016 revision to TSCA gives the EPA more power to determine that chemicals are safe before they can be marketed, it will hopefully provide greater incentives for companies to explore the toxic properties of their products early in the process of development.

Another example of regulatory influences comes from the state of Massachusetts, which passed a Toxic Use Reduction Act (TURA) in 1989. This legislation required companies that used large quantities of potentially hazardous chemicals to investigate potentially safer alternatives to those products. It also created a Toxic Use Reduction Institute to help develop alternatives to hazardous chemicals. The goal of the TURA was to give polluting industries less reason to defend the safety of worrisome chemicals and more reason to explore potential alternatives.

So far, the examples discussed here have focused primarily on strategies for influencing *privately* funded research. We can also engage with the institutions and policies that guide *federally* funded research. One approach along these lines is to influence the overall amount of money provided by the government for research in particular areas. For example, some people worry that the overall proportion of money from the federal government for research and development (R&D) in the United States has shifted dramatically over the past 50 years. Whereas roughly two-thirds of funding for R&D came from the government in the 1960s and one-third came from private industry, the proportion has now flipped so that the government provides only about one-third of R&D funding. As we saw in chapter 5, this can be a source of concern in areas of biomedical or environmental research where industry has significant incentives to minimize any evidence that its products have harmful side effects. Assuming that scientists working on these topics are more likely to respect the values of the public when they obtain their funding from government sources, it is important to ensure that there is adequate public funding for these areas of research.

Another way to engage with the policies for federal research funding is to influence how research projects are chosen. For example, in addition to performing high-profile demonstrations and protests, the AIDS activists described earlier in this chapter agitated to gain positions on the NIH advisory boards that allocated grant funding to specific research projects. This provided a precedent for other groups, such as breast cancer activists, to serve on advisory committees for federal grants so that they could influence decisions about research funding. These strategies are not without their drawbacks, of course. One might worry that high-profile diseases like AIDS and breast cancer could end up receiving disproportionate attention relative to other diseases because of the power of the citizen groups that advocate for research to address them. But these worries highlight the benefits of developing more sophisticated approaches to engagement, not for eliminating engagement altogether.

An additional avenue for influencing funding decisions is to change the criteria used for choosing which grants to fund. For example, beginning in 1997, the NSF decided that one of the two main criteria for evaluating grant proposals should be to look at their "broader impacts." While this criterion has not always been taken as seriously as the other criterion (namely, intellectual merit), it can provide an explicit opportunity to examine the values that inform research proposals. Because competition for federal grant money is so fierce, efforts to influence how the broader impacts criterion is interpreted can have a significant influence on the sorts of research projects that ultimately receive funding.

One can also influence federal research policies by advocating for the funding of specific programs that facilitate reflection on the role of values in science. In the discussion of the NNI earlier in this chapter, we saw that money was set aside to fund grant proposals on the Ethical, Legal, and Social Implications (ELSI) of nanotechnology. This was partly inspired by an earlier ELSI program created by the NIH in response to the Human Genome Project, an international research effort in the 1990s to determine the sequence of human DNA. These ELSI programs provided explicit opportunities for scholars to examine the variety of ways in which values were relevant to these research projects.

The Community Outreach and Education Programs (COEPs) created by the NIH provide one more illustration of federally funded programs that provide avenues for addressing values in science. The National Institute of Environmental Health Sciences (NIEHS) funds a number of research centers at universities, and in 1995 the NIEHS began to require that these centers incorporate COEPs. Initially, these programs were focused largely on educating the public about environmental health hazards, but over time they have also served as an avenue for incorporating public input into the research performed by the centers. Thus, another institutional way to help address values in science is to encourage federal agencies to fund more programs that facilitate community-based participatory research. An even more innovative approach would be for government agencies like the NIEHS to provide more money to fund community-owned and community-managed research of the sort pursued by the West End Revitalization Association. This would give citizen groups even more power to pursue research that serves their interests while contracting with scientists to help them achieve their goals.

CONCLUSION

As the COEPs funded by the NIEHS illustrate, the four approaches to engagement discussed throughout this chapter (see table 7.1) intersect with each other in numerous ways. For example, bottom-up approaches to engagement (category one) are assisted greatly when institutions (category four) are structured to facilitate grass-roots engagement. This can happen through avenues like COEPs or through the inclusion of citizens on advisory panels or through the design of patent policies that allow ethical considerations to play a role in decision-making. But this is a two-way interaction. Activism by citizen groups, as occurred in the AIDS case, is crucial for changing institutional policies and laws to make them more responsive to public values.

There are many other connections between these four approaches. We have already seen that bottom-up, top-down, and interdisciplinary engagement (the first three categories) can be directed toward changing laws, institutions, and policies (category four). Changes to laws and policies can, in turn, facilitate

Table 7.1 AN OVERVIEW OF FOUR MAJOR TYPES OF ENGAGEMENT
THAT CAN HELP TO PROMOTE THOUGHTFUL ANALYSIS
OF VALUES IN SCIENTIFIC RESEARCH

Four Categories of Engagement	Examples
Bottom-up: citizens engaged with scientists	• Oppositional efforts to alter research or to promote more research funding for citizen concerns • Community-based participatory research efforts (CBPR) • Citizens serving on advisory committees
Top-down: formal engagement exercises	• Workshops, consensus conferences, and the National Citizens Technology Forum • "Material engagement" and exhibits at science museums • World Wide Views project
Diverse, interdisciplinary engagement between scholars	• Scholars who critique research that incorporates questionable values • Interdisciplinary collaborations and the Toolbox Project • Embedding humanists or social scientists in labs (the STIR program) • Promoting diversity in the scientific community
Engagement with institutions, laws, policies	• Allowing ethical considerations in the patenting process • Creating a health impact fund or advance market commitments • Altering toxic-chemical regulations • Increasing federal funding for research • Tweaking the broader impacts criterion for grant proposals • Funding ELSI research or CBPR

or hinder the development of formal engagement efforts (category two) and diverse, interdisciplinary research collaborations (category three) depending on the amount of funding allocated for these initiatives. Moreover, diverse and interdisciplinary scholarly communities (category three) are often more open to incorporating input from citizens in their research projects (category one). Either bottom-up or top-down citizen engagement efforts (categories one and two) can also highlight the need for more diverse, interdisciplinary research projects (category three) to address public concerns.

While these four approaches to engagement provide exciting opportunities for altering the values that influence scientific research, it is important to recognize that they are not a panacea. A number of factors make it very difficult to alter the scientific enterprise. Because much of science involves highly specialized knowledge, it can be difficult for citizens and scholars from other disciplines to develop the expertise they need in order to challenge areas of science that they find questionable. It can also be difficult to anticipate in advance how science and technology will shape society. By the time citizens and policymakers realize that a particular body of research has worrisome consequences, it may be too late to alter the course of research significantly. Moreover, when research is being privately funded, concerned citizens are limited in their ability to influence it. Thus, one of the central lessons of this book is that we must continue to explore new and creative ways to steer scientific research in accordance with our values.

SOURCES

For more information about ACT UP and AIDS activism, see the capsule histories available at the website of ACT UP New York (http://www.actupny.org), as well as Crimp (2011), Deparle (1990), and "Police Arrest AIDS Protesters" (1988). Epstein (1996) provides an excellent analysis of the ways that AIDS activists influenced scientific research and policy. Ottinger (2010) discusses the Louisiana Bucket Brigade. WERA and its experiment with community-owned and community-managed research are discussed in Heaney et al. (2007). Kinchy (2010) describes the concept of an epistemic boomerang. For information about community-based research efforts and local knowledge, see Corburn (2005) and Irwin (1995).

Crichton (2002) provides his speculative vision of nanotechnology. For an overview of nanotechnology and the social issues that it raises, see Allhoff et al. (2007). Guston (2008, 2014) and Davies et al. (2009) provide overviews of several public engagement exercises employed in response to nanotechnology, and Irwin (2001) highlights some of the challenges involved in pursuing public engagement.

For more information about Shrader-Frechette's work, see her books *Taking Action, Saving Lives* (2007) and *Tainted* (2014). To learn more about Nancy Tuana's work, see Schienke et al. (2011), Tuana (2010), and Tuana et al. (2012). Erik Fisher's efforts at midstream modulation are described in Schuurbiers and Fisher (2009). The Toolbox Project is described in O'Rourke and Crowley (2013). Fehr (2011) and Wylie (1996) describe efforts by female scientists to challenge value-laden assumptions, methodologies, and theories. Harding (2015) also emphasizes the importance of fostering diversity within the scientific community.

Parthasarathy (2007, 2011) provide analyses of the different patent regimes in the United States and in Europe and their implications for allowing values to influence patent decisions. Information about the Oncomouse is available in Cook (2002) and Park (2004). Biddle (2014a) provides an insightful discussion of the ways that patent policies can suppress research. AMCs are discussed in Berndt and Hurvitz (2005) and Sonderholm (2010). Cranor (2011) provides his discussion of TSCA, and Tickner (1999) reports on the Toxic Use Reduction Act. For information about how breast cancer activists and other citizen groups have served on federal advisory bodies, see Kaime et al. (2010). Information about the NSF's broader impacts criterion is in Frodeman et al. (2013) and Lok (2010).

CHAPTER 8

Conclusion

A Tapestry of Values

On February 2, 1970, two months before the first Earth Day, *Time* magazine published an issue with a special focus on the environment. The magazine put biologist Barry Commoner on the cover and labeled him the "Paul Revere" of ecology. Commoner had become famous because of his political activism against nuclear weapons testing. He was particularly concerned about the potential health effects associated with radioactive fallout from nuclear tests, and he engaged in an extensive campaign to inform the public about its dangers. His efforts played a significant role in the signing of the Nuclear Test Ban Treaty in 1963. Commoner's story is important partly because of his central role in the development of the environmental movement, but his story is also valuable because it provides a counterpoint to the story of Lysenko and Stalin, with which this book began. Like them, Commoner held strong political views and was deeply suspicious of capitalism. But whereas they wielded their political values in problematic ways, Commoner's political values influenced his research in appropriate ways that enriched society. Thus, it drives home this book's argument that what makes values legitimate is not that they are of a particular sort (e.g., conservative or liberal, religious or secular) but that they are incorporated in a transparent fashion, with adequate discussion about whether they meet our ethical and social priorities while doing justice to the empirical evidence.

Building on Commoner's story, this concluding chapter synthesizes the book's major lessons about how values can legitimately influence scientific reasoning. First, it explores how the metaphor of a tapestry can represent many of the ways in which values intersect with research. Second, it highlights some of the major ramifications of the book for our views about science policy and the relationship between science and society. Finally, it addresses a few of the lingering objections that might still be troubling some readers.

BARRY COMMONER

Born in 1917, Commoner was the son of Jewish immigrants from Russia. He grew up in Brooklyn, New York, and attended Columbia University as an undergraduate and Harvard University as a graduate student. During the 1930s, he was influenced by Socialist and Communist rallies, and as an undergraduate he committed himself to engaging in "activities that properly integrated science into public life."[1] For him, one of the most important ways of performing this integration was to inform citizens about important social issues to facilitate their engagement in effective democratic decision-making. This is why he was so committed to providing the public with information about radioactive fallout; he was deeply committed to the notion that the public should be able to decide whether the national security benefits of nuclear testing were sufficient to outweigh the potential health risks.

Commoner's work on radioactive fallout illustrates many of the ways in which values can appropriately influence scientific work. First, he chose to pursue this area of research because he thought it was a particularly important societal issue. Second, he was deeply concerned to communicate his findings in an understandable manner that would equip citizens to influence public policy. In fact, he helped found an organization called the Committee for Nuclear Information (CNI) in order to help disseminate scientific findings about radioactive fallout to the public. Third, he employed research methods that were particularly effective at engaging citizens. For example, he spearheaded a major initiative to collect children's baby teeth so they could be tested for their content of radioactive strontium-90. Not only was the Baby Tooth Survey an important source of information, but it also galvanized public interest in the potential health effects associated with nuclear testing. The survey ultimately received more than 200,000 teeth from concerned parents and showed that the amount of strontium-90 in children's teeth increased fourfold in the first half of the 1950s.

Later in his career, Commoner turned to other topics while continuing to perform scientific research in a manner that served important social values. He developed a Center for the Biology of Natural Systems at Washington University in St. Louis and later moved it to Queens College in New York City. In New York, he turned his attention to a new social challenge: handling urban waste. The mayor of New York City wanted to engage in more waste

1. Egan 2007, 20.

incineration, but Commoner worried that the burning of plastics released dioxin, the highly carcinogenic chemical discussed in chapter 5. His Center studied how dioxin could be transmitted many miles from its initial source and taken up in the food chain. Importantly, he worked to inform local activists about the available science on dioxin and other issues so that they could be better equipped to fight pollution. In a symposium to celebrate his eightieth birthday in 1997, one of the speakers labeled Commoner "the father of grassroots environmentalism."[2] Thus, his work provides an early example of the sorts of engagement between citizens and scientists discussed in chapter 7. Historian Michael Egan sums up Commoner's career well by stating that he "sought to reconnect professionalized science with the public interest."[3] As a result, he exemplifies the message of this book: scientists can incorporate values in their work in many different ways without sacrificing the quality of their science.

A TAPESTRY OF VALUES

As we saw in the preceding chapters, and as table 8.1 summarizes, values can play a wide range of legitimate roles in scientific reasoning. By reflecting on the cases considered throughout the book, we can develop a more insightful vision of how values intersect with science. The metaphor of a tapestry provides a helpful way of summarizing many features of these cases. According to this metaphor, scientific reasoning can be thought of as a tapestry consisting of numerous components or "threads" that scientists need to bring together in order to draw conclusions. Some of these threads are relatively rule-governed and free of values, such as logical principles and mathematical techniques. But these aspects of science are intertwined with value-laden threads, such as background assumptions, terminological choices, and decisions about what methods are most appropriate to employ. The tapestry metaphor highlights at least three important lessons about scientific practice: (1) the analytical or rule-governed components of scientific reasoning are deeply interwoven with components that are influenced by values; (2) the roles of values can be analytically disentangled from other aspects of scientific reasoning; and (3) specific influences of values can have "ripple" effects across science that require careful attention.

First, just as the threads of a tapestry are tightly interwoven, we have seen that scientific practice is deeply intertwined with value judgments. Most people would already acknowledge that values play various roles along the "sidelines"

2. Egan 2007, 195.
3. Egan 2007, 197.

Table 8.1 A LIST OF MAJOR ROLES THAT VALUES CAN LEGITIMATELY PLAY IN SCIENTIFIC PRACTICE, ALONG WITH SIGNIFICANT EXAMPLES PROVIDED IN THIS BOOK

Roles for Values	Examples
Choosing topics to study	• Deciding whether or not to pursue research concerning variations in intellectual abilities based on sex or race • Choosing how to allocate federal research funding, including the role of Congress in the process • Evaluating how well privately funded pharmaceutical research meets social priorities
Deciding how to study a particular topic	• Choosing the best methods for research on agricultural innovations • Choosing assumptions in risk assessments of environmental pollutants • Identifying specific questions to explore in research on depression or cancer
Determining the aims of scientific inquiry in a particular context	• Determining which qualities of risk-assessment methods to prioritize (e.g., speed or accuracy) • Deciding which theories to pursue in anthropology • Choosing which predictive capacities of climate models to prioritize
Determining how best to respond to uncertainty	• Deciding how boldly to communicate conclusions about climate change or endocrine disruption • Deciding how much evidence to demand in order to draw conclusions about toxic substances like dioxin
Deciding how to describe, frame, and communicate results	• Identifying the most appropriate ways of framing research on genetic factors that influence behavior • Choosing terms and metaphors for describing invasive species or environmental phenomena • Evaluating categories for racial classifications or environmental regulations

of scientific practice: guiding the choice of research topics, influencing how scientists treat each other and their experimental subjects, and determining how science is used in society. But the cases that we have studied illustrate that values have much more integral roles to play in scientific reasoning. They can influence the specific questions that scientists ask, the methods they use for studying them, the background assumptions they employ, the sorts of models they develop, the amount of evidence they demand, and the language they use for describing their findings. Thus, scientific reasoning is thoroughly imbued with value influences. And as philosopher Heather Douglas has persuasively argued, scientists have ethical responsibilities to consider the effects of their value-laden choices on society.

A report by the US National Academy of Sciences (NAS) makes essentially the same point as the tapestry metaphor when it describes the practice of characterizing risks as an "analytic-deliberative" process. When researchers attempt to assess risks, they employ a wide variety of technical methods that the report describes as "analytic." But deciding which technical methods to use and how to interpret the results requires a host of value judgments. In accordance with the seventh chapter of this book, the NAS concludes that these value judgments are best handled by bringing together a broad range of stakeholders to deliberate about the best approaches for handling them. Thus, risk characterization is described in the report as a procedure in which technical "analysis" and "deliberation" about value judgments are intertwined to form an "analytic-deliberative" process. The term "co-production" has also been used by scholars to describe this intertwining of social values with scientific practice.

In addition to highlighting how scientific practice is interwoven with values, the tapestry metaphor also illustrates how specific roles for values can be analytically disentangled from other elements of scientific reasoning. Even though the threads of a tapestry are tightly interwoven, one could (with enough effort) distinguish different threads from each other. Similarly, we have seen throughout this book that it is possible to clarify a variety of different ways in which values influence scientific practice. Saying that values can be "analytically disentangled" from other elements of science means that even though we cannot remove the values, we can identify specific value influences and distinguish them from other aspects of scientific reasoning.

This is an important lesson because it means that we can think critically about the specific ways that values are influencing science. We do not have to make the blanket claim that it is always legitimate or illegitimate for values to influence scientific reasoning. Rather, we can examine how values are influencing science in particular cases and consider whether those influences are legitimate. Throughout the book, we encountered two major justifications for consciously bringing values into specific aspects of scientific reasoning. First, it is sometimes virtually unavoidable that scientists will have to make choices that serve some values over others, so scientists have responsibilities to make these choices thoughtfully rather than carelessly. Second, in some cases, values help scientists to achieve legitimate goals associated with serving society. We also encountered three conditions (transparency, representativeness, and engagement) that are important for deciding which values should play a role in science. These conditions will be discussed further in the next section of this chapter.

In a third way, the tapestry metaphor helps to illuminate the role of values in science by illustrating the "ripple" effects that values can generate and the careful attention that is required to recognize those effects. When a single thread of a tapestry is pulled, it can warp the surrounding fabric and alter the tapestry's image in unexpected ways. In the same way, it can sometimes be difficult to discern whether particular influences of values are appropriate or not because they can have complex influences on science that are difficult to recognize. In some cases, seemingly problematic influences can turn out to be acceptable, and in other cases seemingly innocuous influences can turn out to be problematic. Thus, while this book has tried to suggest strategies for distinguishing whether particular value influences are appropriate, it is wise to remember that these distinctions are not always easy to make.

For example, chapter 4 argued that it can be appropriate for scientists to pursue theories that interest them or that promote their individual values, such as when anthropologists purposely developed and explored theories of human evolutionary development that emphasized the roles of women. However, it may not always be beneficial for scientists to focus on their individual values when deciding which theories to pursue. Suppose there were not much diversity among the scientists working in a particular field. Encouraging them to pursue theories that promoted their personal values might not move the field forward in a very effective or socially responsible way because they might not generate a very diverse range of theories to test and compare.

Consider also the discussion of the pharmaceutical industry in chapter 3. We saw that the problematic influences on pharmaceutical research are not all blatant and obvious. For example, in their efforts to promote the scientific legitimacy of their field, psychiatrists in the 1970s and 1980s focused more of their attention on biological explanations for mental health problems (as opposed to social or psychological explanations). In conjunction with other factors, such as the pharmaceutical industry's desire to emphasize patentable treatments, this may have resulted in an overemphasis on drug treatments for depression as opposed to other potential treatments. In principle, there does not seem to be anything wrong with physicians pursuing greater scientific rigor for their field or companies seeking to develop products that can generate a profit, but in this case the convergence of values may have resulted in suboptimal treatments. Thus, the tapestry metaphor reminds us that even when particular roles for values are typically legitimate (or illegitimate), it is still important to explore their wider effects.

RAMIFICATIONS FOR SCIENCE POLICY

If we take seriously the lessons of this book, including the tapestry metaphor, it requires rethinking some common views about science that have been prominent in recent decades. For example, as discussed in chapter 2, Vannevar Bush published a highly influential report, *Science: The Endless Frontier*, that had a major impact on science policy after World War II. He proposed creating an agency (which ultimately became the National Science Foundation) that would provide research funding for university scientists while providing them with maximum autonomy to pursue their interests without outside interference. Bush contended that basic research was a crucial national resource that provided the fuel to generate subsequent innovations in support of national defense and economic growth. Bush's work contributed to the development of several concepts that have had a major influence on subsequent science policy.

Previous Science Policy Concepts

Bush's idea that basic research leads straightforwardly to applied science, which in turn consistently generates technological developments and social benefits, came to be called the "linear model." Closely related to this model was the notion that scientists have a "social contract" with society, according to which they should be given maximum autonomy to pursue basic research as they see fit. Those who espouse the notion of a social contract contend that when science is conducted autonomously, it will flourish and ultimately generate extensive social benefits. These ideas tended to support the notion that value judgments could be largely ignored or excluded from basic research and considered only in the final stages when science gives rise to new commercial products or guidance for policymakers.

Twentieth-century science policy also tended to use a "deficit" model to characterize the public's understanding of science. According to the deficit model, public opposition to new developments in science and technology is typically caused by people's lack of scientific knowledge. For example, on this view, people's suspicions about genetically modified foods or vaccines or nuclear power can typically be attributed to their ignorance about the scientific evidence related to these topics. The deficit model resonates well with the linear and social contract models because it suggests that the public does not know enough about science and technology to influence the course of research in an intelligent fashion. Instead, it is best to let scientists make their own decisions about how to do their work, and (according to the social contract) it will ultimately work out in society's best interest.

According to science-policy scholar Roger Pielke, another common tendency has been for people to assume that scientists can provide relatively value-free advice to guide policymakers. According to Pielke, there is an "Iron Triangle" of three groups who all find it appealing to maintain this assumption. First, scientists like to strive for value freedom because it fits with their traditions and promotes their authority. Second, politicians like the idea that they can avoid making difficult decisions and instead let the value-free authority of science guide public policy. Third, special-interest groups like to maintain the value-free reputation of science because they can then wield their preferred scientific ideas with more authority to advance their social goals.

The notion that scientific reasoning is a tapestry, interwoven with value judgments, complicates all these traditional ideas about science and its relationship to public policy. Most obviously, it thwarts the Iron Triangle of stakeholders who would like to see science as a value-free source of guidance for policymakers. But it also complicates the linear, social contract, and deficit models. The linear model suggests that value judgments can be largely ignored until the final stages of research, when basic research finally gives rise to new products and guidance for policymakers. We have seen, however, that value judgments play an important role throughout the course of scientific research. The social-contract model also seems misguided because it is doubtful that the best way to handle all these value judgments is simply to leave scientists to their own devices, without any outside input. And the major assumption of the deficit model—that public opposition to science stems primarily from ignorance—also loses much of its force when science is recognized to be a tapestry of values. In many cases, public opposition may stem from disagreements about the values underlying specific research projects rather than from ignorance.

A Path Forward

If these ways of thinking about the relationships between science and society are inadequate, then we need to develop new approaches. The cases that we have examined throughout this book suggest that as we navigate the relationships between science and society, we should be focusing on at least three conditions (engagement, transparency, and representativeness) that are typically important for incorporating values into science in a fully justifiable manner. Chapter 7 emphasized that *engagement* is central to addressing value judgments in science in a responsible way. It discussed bottom-up engagement associated with citizen groups, top-down engagement created through formal efforts to elicit public input, engagement among diverse and interdisciplinary groups of scholars, and engagement with institutional rules and policies. All these forms of engagement provide avenues for stakeholders to incorporate their perspectives in the tapestry of values that make up scientific research. In this way, individual scientists are no longer left to make value judgments on their own (as the social contract model suggested), but rather a range of different stakeholders can deliberate about how best to address these value judgments.

Another central condition that emerged throughout many of the preceding chapters was *transparency*. We found in a number of cases that it was virtually impossible to avoid making value judgments when performing scientific research. In other words, even if scientists did not intend to support some values over others, they could not avoid making choices that did so. For example, chapter 3 described how scientists are forced to make a variety of assumptions when assessing the risks associated with environmental pollutants. The available evidence typically does not determine which assumptions are best, but some end up being more supportive of public health, whereas others are more favorable to the chemical industry. Thus, there is little hope of addressing these assumptions in a manner that is value free; the best we can do is to be transparent about our assumptions so that others can take them into account. Similarly, at the end of chapter 5, we saw that one of the crucial distinctions between those who manufacture scientific uncertainty inappropriately and those who raise legitimate questions about science is that those who act inappropriately often fail to be fully transparent about their knowledge or motivations. The same emphasis on transparency also appeared in chapters 4 and 6, when we found that scientists should strive for transparency about their cognitive attitudes toward theories and about the role of values in their descriptions of scientific information. Admittedly, it is somewhat impractical for scientists to be fully transparent about their values, but recent efforts to make scientists provide more information about their data and methods can enable others to scrutinize their work and recognize how it supports some values over others.

A third condition that emerged throughout the preceding chapters was *representativeness*, meaning that the values that influence research should represent our ethical and social priorities. This idea was highlighted in chapter 2, when we encountered the worry that our congressional representatives might not represent the overall concerns of their constituents very well. In the same chapter, it was even more obvious that the pharmaceutical industry's research priorities do not represent all the needs of the world's citizens well. Diseases that primarily afflict low-income nations do not receive much research, and even drugs produced for the citizens of wealthy countries are often not particularly innovative. The same concern arose in chapter 5, when we saw that those who have attempted to manufacture doubt about the harmful effects of tobacco and industrial chemicals and climate change typically represent the concerns of a few wealthy corporations rather than the interests of the broader public. Admittedly, it can be difficult to weigh the differing values in society and decide what mixture of values is truly representative. Nevertheless, the examples in chapters 2 and 5 illustrate that there are some situations in which it is not only the case that research efforts focus on a limited group of powerful interests, but strong ethical reasons also support moving research in other directions. It is still typically not easy to shift the research enterprise so that it represents our ethical commitments or the values of a broader array of stakeholders, but many of the engagement efforts described in chapter 7 are designed to help achieve this goal.

Fortunately, the goals of promoting engagement, transparency, and representativeness tend to coincide. For example, efforts at engagement are often motivated by the desire to remedy failures in transparency or representativeness. As discussed in chapter 7, when AIDS patients felt that the policies of the federal government and the pharmaceutical companies were not representing their values, they formed ACT UP in order to advocate for policies that better served their values. Again in chapter 7, we saw that the engagement efforts of interdisciplinary scholars like Kristin Shrader-Frechette are often directed toward making questionable research assumptions more transparent. Improving transparency and representativeness can also contribute to better engagement. For example, a central theme of chapter 6 was that scientists should strive for greater transparency about values as they communicate scientific information so that members of the public can better understand whether or not their values are being served. Thus, as we take steps to achieve each of these three goals (engagement, transparency, and representativeness), we will hopefully end up promoting all three in the end.

OBJECTIONS

Before concluding, it may be helpful to respond to several objections (see table 8.2). First, one might wonder, if it is acceptable for science to be saturated with values in the ways this book has described, how could the value-free ideal for science have become so prominent? Given the widespread acceptance of the value-free ideal, it seems that there must be some legitimacy to it. My response is that there is indeed a kernel of truth in the traditional notion that science should be value-free. Like all scholars, scientists should indeed try to be fair-minded and objective in evaluating the evidence available to them. They should not fall prey to the problem of wishful thinking and accept conclusions merely because they want those conclusions to be true. Rather, scientists should demand careful reasoning and appropriate evidence in support of their conclusions. But we can acknowledge all these limitations while recognizing that values still have many legitimate roles to play in science.

Perhaps the reason that so many people insist that science should be value-free is that they have a particular picture of what it would mean for values to be incorporated into science. They probably envision scientists deliberately ignoring some pieces of evidence and misinterpreting other pieces of evidence so they can draw conclusions that serve their social, political, or religious goals. But in reality, as we have seen throughout this book, values can be incorporated into science in a much more thoughtful and appropriate manner. In many cases, scientists are already making particular assumptions or choosing

Table 8.2 POSSIBLE OBJECTIONS TO THE BOOK'S LESSONS, TOGETHER WITH RESPONSES

Objections	Responses
• Given the widespread acceptance of the notion that science should be value-free, there must be some legitimacy to this idea	• Scientists should indeed be fair-minded and avoid wishful thinking, but the value-free ideal may rest largely on misinterpretations of what it means to incorporate values in science
• Acknowledging that science is value-laden destroys its authority as a source of knowledge that everyone can accept	• In most cutting-edge research, there is plenty of room for disagreement, so it is best to acknowledge this disagreement and explore the potential causes for it
• Not all areas of science are as value-laden as the examples in this book	• It is still wise to expect that values may be relevant and to provide opportunities for engagement to identify potential value influences
• The positive vision expressed in this book is too demanding of scientists, citizens, and societal leaders	• A substantial amount of progress can be made by strengthening efforts that are already present

particular terms or using particular methods that result in some values being supported rather than others. Thus, to acknowledge the value-laden nature of science is to recognize that these choices should be made in a thoughtful and transparent fashion, with explicit attention to their connections with our values. In many cases, the choice is not whether to incorporate values in science or to remain neutral; rather, the choice is whether to make value-laden choices transparently or whether to do so without recognizing their significance.

A second worry is that if values are allowed to play a role in science, the unique role that science is expected to play in society will be sacrificed. Whereas people disagree about politics and ethics and religion, science is expected to be a source of information that everyone can accept and use as a starting point for making decisions as a society. But the problem with this objection is that it does not describe cutting-edge research very well. While some areas of science do eventually become fairly settled (one thinks, for example, of the basic principles of physics or chemistry), much of science is not like this. When researchers are studying the biological factors that influence human behavior, or the health effects of toxic chemicals, or the best ways of developing new agricultural technologies, there is plenty of room for disagreement. Thus, in these research situations there is little to be lost and much to be gained by acknowledging the role of values. Both experts and ordinary citizens already disagree about which conclusions are most reliable and how best to interpret the available evidence, so scientific objectivity would actually be promoted by being more explicit about the reasons for these disagreements.

A third complaint might be that not all areas of science are as value-laden as the examples discussed throughout this book. Critical readers might notice that the chapters are full of examples from complex and socially significant areas of science like human biology, medicine, risk assessment, climate change, anthropology, agriculture, and toxicology. The critics might point out that values are less likely to play a significant role in fields like physics and chemistry. This is probably true, but it does not mean that the lessons of this book are completely irrelevant to these fields. Important questions still need to be asked regarding what topics to study, how to study them, how much evidence to demand before drawing conclusions, which phenomena are most important to model, and how to communicate the resulting findings. And even if values play a relatively minimal role in some areas of science, values obviously have a central role to play in a wide array of other scientific fields. Thus, it is likely better for scientists to assume that values could be relevant to their work and to welcome opportunities for engagement rather than assuming that values have no role to play in their areas of expertise.

A fourth objection might be that the positive vision proposed in this book—focusing on engagement, transparency, and representativeness—asks too much of everyone involved. For scientists, it appears to require an enormous amount of work in order to reflect on the social ramifications of the choices that they make in their daily practice. For citizens, it seems to require that they spend an unrealistic amount of time reflecting on the state of contemporary science and the ways in which it does or does not serve their interests. And for government policymakers and corporate leaders, it seems to require that they take very difficult steps to steer their research portfolios in ways that serve the public interest.

In some respects, the vision laid out in this book will indeed be difficult to achieve, but a substantial amount of progress can be made merely by maintaining and strengthening a number of steps that are already under way. For example, scientists do not have to spend all their time thinking about the social ramifications of their research in order to make responsible decisions about the values that inform their work. They can make significant progress by being responsive to funding agencies and concerned citizens, by encouraging greater diversity among the scholars in their fields, and by developing interdisciplinary collaborations when opportunities arise. Similarly, citizens can influence science in positive ways even by taking relatively modest steps. For example, we have seen that many people are already motivated to participate in citizen groups that promote agendas they care about. Others can donate money to NGOs or engage in occasional political activities. Admittedly, attempting to change government policies and corporate behavior remains a difficult task. But the difficulty of the task is hardly a reason not to try; to do otherwise is to remain resigned to a scientific research agenda that fails to meet our ethical and social priorities.

CONCLUSION

Philosopher Sheldon Krimsky has argued that Barry Commoner's research provides an excellent example of what he calls "public-interest science," namely, science geared toward addressing pressing social issues like inequality, public health, and environmental degradation. Another scientist highlighted by Krimsky for doing public-interest work is Luz Claudio, a scientist at the Mount Sinai School of Medicine. Not only does her research address important public needs, but it also illustrates many of the lessons emphasized throughout this book. In a study performed during the late 1990s, she found that the impact of asthma on communities in New York City varied dramatically depending on their wealth. In a poor area of East Harlem, for example, there were over 200 hospitalizations per 10,000 people for complications related to asthma each year, whereas in some wealthy areas of Manhattan and Queens, the hospitalization rate per 10,000 people was zero. Overall, the researchers found that hospitalization rates from asthma were more than 20 times higher in low-income areas than in more affluent ones. These results supported a growing realization that low-income people and minorities were suffering disproportionate impacts from urban air pollution. Other studies found that the presence of asthma among children in the Bronx was twice the national average and that asthma hospitalization rates for minorities in New York were seven times higher than the rates for whites.

One way in which Claudio's research illustrates the book's lessons is by showing how engagement can drive research in ways that serve community values. When Claudio initially set out to pursue research of relevance to disadvantaged communities in New York City, she did not realize that asthma was a major concern for them. Based on a grant from the Environmental Protection Agency, however, Claudio sat down with community leaders from Harlem, Brooklyn, and the Bronx to talk about their priorities. When she found out that asthma was a high priority, she worked with community leaders to develop the study discussed above. She later collaborated with community leaders to examine how increased power generation at a plant on the lower East Side of New York would affect local citizens. The results empowered community members to negotiate with the power company to alleviate toxic emissions.

Claudio's research illustrates many of the forms of engagement discussed in chapter 7. First, she grew up in Puerto Rico and was deeply influenced by a grandmother whom she described as an indigenous "curer" or healer.[4] Thus, she illustrates the benefits of promoting a diverse scientific workforce that can bring unique perspectives to the research teams with which they work. But she was also empowered by institutional policies that encouraged researchers to bring community input into their research projects. As chapter 7 noted, the National Institute of Environmental Health Sciences (NIEHS) created Community Outreach and Education Programs (COEPs) within a number of the centers it funded during the 1990s. Claudio served as director of the COEP at Mount Sinai, which placed her in an ideal situation to perform research that was informed by local community values. Finally, Claudio's work with the residents of the lower East Side was further strengthened because she was able to collaborate with a citizen group, the East River Environmental Coalition, which was already working to protest local pollution.

Many of the same features displayed by Claudio's research are also present in the work of another New York City COEP, which was based at Columbia University's Mailman School of Public Health. The Columbia COEP developed a close partnership with the community group West Harlem Environmental Action (WE ACT). WE ACT was formed in the late 1980s in response to concerns about air and water pollution. Youth from WE ACT collaborated with scientists from the Columbia COEP to study how air pollution from diesel buses and trucks might be affecting their health, and the resulting information was used to help combat diesel traffic in the community. The scientists from Columbia emphasized that the members of WE ACT helped steer the design and interpretation of these studies, and community members ultimately served as co-authors of the publications. According to Mary Northridge, the director of the COEP at Columbia, "Our high community response rates, culturally sensitive protocols, and cautious interpretation of data are the direct result of our close collaborations with community residents in the design, conduct, and dissemination of our research findings."[5]

The work of the COEPs at Mt. Sinai and Columbia highlights how scientists have become more sophisticated in recent decades at incorporating the public into research. Barry Commoner demonstrated in a vivid fashion how scientists can both pursue research that serves public values and communicate the results in a manner that promotes democratic decision-making. But the COEPs went even further and demonstrated how scientists can even work with citizens to design studies and determine how to interpret the results. In fact, the name for these programs was later changed from Community Outreach and *Education* to Community Outreach and *Engagement*, partly because the term "engagement" better expresses the two-way flow of information to which

4. Krimsky 2003, 192.
5. Claudio 2000, A451.

these programs aspire. The COEPs in New York illustrate the striking outcomes that can be generated when powerful institutions facilitate and encourage the right forms of engagement.

As these COEPs illustrate, scientific research has tremendous potential to shape society for the better. Unfortunately, much of this potential is wasted when scientists do not recognize the ways in which their research choices support some values and detract from others. Throughout this book, we have focused on five research choices that are particularly significant: deciding what topics to study, determining how best to study them, deciding on the aims of the research, responding to uncertainty in the available evidence, and choosing how to communicate the findings. Whether explicitly or implicitly, scientists make value judgments when they decide how best to handle these choices. While proponents of the value-free ideal worry that incorporating values in science will detract from scientific objectivity, we have seen that objectivity can actually be enhanced when implicit value judgments are brought into the open and subjected to thoughtful scrutiny and deliberation. A central theme of this book has been that scientists can legitimately incorporate values into their work if they are sufficiently transparent about them, if the values adequately represent our ethical principles and social priorities, and if appropriate forms of engagement are in place for scrutinizing and steering these values. Through thoughtful engagement between public- and private-sector institutions, citizen groups, scholars, and scientists, we can guide research in ways that serve our values and enrich our lives.

SOURCES

Egan (2007) provides a very helpful study of Barry Commoner's life and work. Further information about Commoner can be found in Krimsky (2003).

Douglas (2003, 2009) argues that scientists have responsibilities to consider the social consequences of their value-laden choices. The NAS report that calls for analytic-deliberative processes is NRC (1996). The language of co-production is found in Harding (2015) and Jasanoff and Wynne (1998). For a discussion of how we can examine the combination of value influences in a particular context and consider whether the overall influences are problematic or not, see Solomon (2001).

For a history of science policy in the United States, see Elliott (2016c). Bush (1945) provides an influential early statement of the linear model. Guston (2000), Pielke (2007), and Sarewitz (1996) are excellent sources for evaluating the linear and social contract models. The deficit model is discussed in Toumey (2006) and Wynne (1992, 2005). Pielke (2007) discusses the Iron Triangle. For further information about engagement, see the readings discussed at the end of chapter 7 as well as Longino (1990, 2002). For the importance of transparency, see Douglas (2009) and Elliott and Resnik (2014). Further perspectives on representativeness are in Elliott (2011b), Intemann (2015), and Kourany (2010).

Further reflections about how the value-free ideal developed are in Douglas (2009) and Proctor (1991). For the notion that disagreement in science should be expected and respected, see Wickson and Wynne (2012). For insightful discussions about areas of science that are particularly laden with values, see Funtowicz and Ravetz (1992). Shrader-Frechette (2007) discusses how citizens can take a variety of actions to influence scientific research in accordance with their values.

The COEP at the Mount Sinai School of Medicine is described in Claudio (1996), Krimsky (2003), and Noble (1999). More information about the COEP at Columbia University can be found in Claudio (2000) and Corburn (2005).

DISCUSSION QUESTIONS

CHAPTER 1

(1) Before you started reading this book, would you have been inclined to say that science should be value-free? If so, why? If not, what roles for values would you have been willing to accept?

(2) Do you think that the comparison between trying to remove knives from kitchens and trying to eliminate all values from science is a fair one? Why or why not?

(3) Based on what you read in chapter 1, do you think that Colborn and her colleagues were influenced by values in appropriate ways? If so, what are the key differences between the ways values influenced them and the ways values influenced figures like Stalin and Lysenko? If not, is your worry that Colborn and her colleagues should not have been influenced by values at all, or is the problem that they sometimes prioritized the wrong values?

(4) Can you think of other historical or current examples where values influenced science in ways that you regard as appropriate or inappropriate?

(5) Do you agree that there are many situations in science where value influences are unavoidable, so it is better to be explicit about value-laden choices rather than trying to avoid making value judgments?

(6) Do you agree that values tend to be appropriate in science when they are incorporated transparently, when they represent major ethical or social priorities, and when there are appropriate forms of engagement to reflect on them? Can you think of other conditions that determine whether values are appropriate or not?

(7) When you hear about citizen involvement in scientific research (as in the Woburn case), do you find it appealing or not? Do you worry that citizens are likely to harm the quality of the research, or do you think their involvement provides an effective way to steer research in more socially beneficial directions? Are there ways of gleaning the benefits of their participation while minimizing potential harmful effects on research?

CHAPTER 2

(1) Do you agree that research into differences in cognitive abilities between gender and racial groups should be a low social priority? Why or why not?

(2) Are there forms of research into gender or racial differences that you regard as a high priority? For example, would you support research on the ways in which women might be socialized in a manner that inhibits their interest or ability in math and science? If so, what is the difference between these forms of research and the research projects that you would attempt to discourage?

(3) Are there areas of research that you think are so potentially harmful to society that you would either ban them or prohibit publication of them? What is it about those areas of research that makes them so problematic?

(4) What role, if any, do you think Congress should play in guiding funding decisions at the National Science Foundation? Would you allow Congress to have a greater influence over research funding decisions in other federal agencies, such as the Department of Defense? Why or why not?

(5) In addition to congressional oversight, can you think of other strategies that could be used to make funding decisions at agencies like the National Science Foundation responsive to taxpayers' interests?

(6) If Representative Smith was indeed engaged in a political effort to suppress research on climate change, were his efforts any different from those who would like to discourage research on gender or racial differences in cognitive abilities? Why or why not?

(7) Do you agree that it is a bad idea to have congressional representatives double-checking decisions by government agencies to fund specific grants? Can you imagine ways to structure the process so that it would not become overly political and/or short-sighted?

(8) Do you agree with the fair-share principle—i.e., the notion that research funds should be allocated toward studying diseases in proportion to the amount of suffering that those diseases cause? Would you agree that this is the ethical thing to do, even if a large portion of research funding in the United States would end up going toward diseases that are not significant problems in the United States?

(9) What solutions do you think are most promising for addressing the problems with contemporary biomedical research? Would you try to alter our current patent system in some way or would you prefer to pursue alternative solutions?

CHAPTER 3

(1) Based on what you learned in chapter 3, would you be inclined to push for extensive research on golden rice, or do you think that other solutions for vitamin A deficiency are more important to pursue?

(2) People often use the story of the blind men and the elephant as an illustration of the way different religions can all share partial understandings of the same spiritual reality. Do you think it is fair to say that science displays some of the same characteristics? Why or why not?

(3) Should we be disappointed that some groups abandoned the IAASTD project, or was it helpful that the report highlighted the strong value differences between different groups? Do you think that the conflict over the report could have been avoided if the organizers had been more strategic and creative?

(4) Chapter 3 suggests that policymakers are in danger of forgetting that there are social and political strategies for helping farmers as well as biological and technical strategies. What are some of the strengths and weaknesses of these different research strategies? Can you think of other social problems that are often approached with technical strategies but that could also be addressed using social or political changes?

(5) The second section of chapter 3 suggests that scientists are sometimes forced to make assumptions before they have adequate information to guide them, and therefore values can legitimately influence them. Do you agree? Can you think of ways to avoid making assumptions under these circumstances?

(6) When citizens challenge experts because they think they are making questionable assumptions, can you think of good ways to determine whether those criticisms are legitimate or whether they reflect ignorance on the part of the citizens?

(7) Suppose that it would be better for society as a whole if we put more effort into studying cancer prevention rather than cancer treatment, or if we put more effort into developing safer chemicals rather than assessing the risks of old ones. How can we achieve more research of this sort? Is it politically feasible to convince governments to do so? Could private foundations or philanthropists fill the gap? Can you think of ways to encourage private industry to fund more of these types of research?

CHAPTER 4

(1) Do you think it is justifiable for regulatory agencies to employ scientific approaches that do not yield the most accurate results? If so, how can we decide when they have "gone too far" in sacrificing reliability for the sake of producing results more quickly or easily?

(2) Chapter 4 considers three cases in which somewhat inaccurate approaches were used by government agencies: the NCD river restoration technique, the CEPA method of risk assessment, and the RAMs used for wetland assessment. What do you think of these cases? Are they all justifiable? Are any of them more or less justifiable than the others? Why or why not?

(3) The second section of chapter 4 argues that it is appropriate for scientists to allow values to influence them when they are deciding which theories to develop or explore. Do you agree, or do you think scientists should focus on pursuing the theories that seem most likely to be true?

(4) The second section of chapter 4 argues that feminist anthropologists could legitimately choose to focus on developing theories of human evolution in which women played a particularly prominent role. Do you agree that this is appropriate? Why or why not? Would it be equally appropriate for anthropologists who want to diminish women's roles in society to focus on exploring theories of human evolution in which women did not play an important role? Why or why not?

(5) Chapter 4 provides an explanation for why Willie Soon's pursuit of a solar radiation theory of climate change was unacceptable. Do you agree with that explanation? Are there other ways in which he could be criticized?

(6) The third section of chapter 4 shows that, in at least some cases, scientists need to choose what features of the world are most important to represent when they develop models or theories. Do you think that scientists always need to make these decisions about what to represent when they develop models or theories, or do these choices just occur sometimes? Does this mean that values are always relevant to developing models or theories, or just sometimes?

CHAPTER 5

(1) Do you think it was appropriate for James Hansen to claim that climate change was occurring, even though the evidence was somewhat unclear? Are there ways he could have altered his message to make it more appropriate? Do you think there are any important differences between the Hansen, Colborn, and Kangas cases?

(2) Table 5.1 presents three idealized approaches to science communication that scientists could adopt. Which, if any, of these approaches do you find to be most justifiable, and why?

(3) Can you think of strategies that scientists could use when communicating with the media in order to prevent the sorts of public panics that occurred in the Alar and vaccine controversies?

(4) Do you agree with Douglas that scientists should allow values to influence the amount of evidence that they demand when they are drawing conclusions or interpreting evidence or making assumptions? Does your answer depend on the specific type of decision that scientists are making—for example, setting statistical significance levels or interpreting evidence (like rat tissue slides) or making assumptions?

(5) Do you agree that we would have more success resolving social controversies about topics like climate change, evolution, vaccines, and GMOs

if we focused more on resolving disagreements about values rather than on "hitting people over the head" with scientific information? Does this sacrifice the respect that we should be giving to science?

(6) Do you agree that the tobacco industry and the fossil fuel industry acted inappropriately when they challenged the scientific evidence that conflicted with their financial interests? What would have been the responsible course of action for these companies? Do you think chapter 5 is too critical of industry?

(7) The last section of chapter 5 focuses on examples where values seemed to generate uncertainty in an inappropriate fashion. Can you think of examples where it was helpful and legitimate for values to promote scientific uncertainty or controversy?

CHAPTER 6

(1) What range of goals do you think responsible scientists should be trying to achieve when they engage in communication? Should they strive only to be as accurate as possible? Should they incorporate one or more of the other goals discussed at the beginning and end of chapter 6? Why or why not?

(2) Do you agree that scientists invariably frame the information they provide? Do you agree that one good way of handling frames is to become aware of the options and to backtrack when using controversial ones? Can you think of other ways to handle them responsibly?

(3) The first section of chapter 6 discusses a number of examples of different frames (e.g., social progress, scientific uncertainty, teach-the-controversy, economic competitiveness, etc.). Can you think of other major frames that scientists use when communicating with the public?

(4) Larson argues that scientists should try to employ metaphors that promote socially beneficial values, like environmental sustainability. Do you think he is being too aggressive about incorporating values into science? Would you feel more comfortable if scientists tried to employ metaphors that did not have very strong values associated with them? Is that feasible?

(5) Larson provided an extensive list of metaphors from the environmental sciences. Try to think of metaphors from other areas of science, such as chemistry or physics. Do you think these metaphors incorporate any values?

(6) Do you agree that in many cases scientists cannot find entirely "neutral" terminology, meaning they are forced to choose terms that subtly support some values over others?

(7) Based on the examples provided in chapter 6 or other examples that you can think of, can you develop a list of ways in which the choice of scientific

terms can influence society? For example, different terms might do a better or worse job of drawing people's attention to a problem, or they might tend to generate positive or negative attitudes. Try to come up with other examples.

(8) Do you think it is a good idea to employ racial categories in medicine? Why or why not? Does it depend on the circumstances? Are there ways of employing these categories that maximize their benefits while minimizing their harms?

CHAPTER 7

(1) Of the different approaches for engagement discussed in chapter 7, which do you like best, and why? Which do you think are most effective? Which are easiest to implement? Are particular approaches best for some purposes but not others?

(2) When you hear about AIDS activists influencing the design of clinical trials, does it worry you or reassure you? Can steps be taken to make sure their involvement does not harm the quality of the trials? Can we decide what counts as a "high-quality" trial without consulting those who will be affected by the results?

(3) Do you think that the sorts of formal engagement exercises discussed in the case of nanotechnology have much potential to influence the future development of this area of research? Which of the formal approaches discussed do you like most and least? Can you think of even better ways to elicit the public's perspectives on this area of technology?

(4) Do you think that philosophers like Kristin Shrader-Frechette or Nancy Tuana are likely to have a more fruitful impact on the values that inform research projects when they collaborate with scientific researchers or when they evaluate scientific projects critically from the perspective of an "outsider"? Does it depend on the situation?

(5) If you had to choose between trying to alter the institutional incentives and policies that guide private-sector research or those that guide public-sector research, which would you focus on changing? What steps would you take to change those incentives or policies?

(6) Can you think of additional ways to incorporate ethical and social values more effectively in scientific research?

CHAPTER 8

(1) Do you accept the tapestry metaphor as a good representation of the roles that values play in science? What do you think are the greatest strengths and weaknesses of the metaphor?

(2) Do you think that the linear, social contract, and deficit models should be abandoned, or do you think some elements of those models are reasonable?

(3) What do you think of the suggestion in chapter 8 that the "path forward" for relating science with society should focus on the concepts of engagement, transparency, and representativeness? Do you think that any of these concepts are more important than the others? Which one is easiest or most difficult to achieve?

(4) The end of chapter 8 suggests several objections that could be raised against the main themes of the book. Of these objections, which do you find to be most compelling? Do you think the response in the book is adequate? Are there other objections that you think need to be addressed?

(5) Chapter 8 begins with the story of Barry Commoner, and it ends with the story of the COEPs in New York City. Do you find the approach to science exemplified by the COEPs and by Commoner to be appealing? Do you have any concerns about their activities?

REFERENCES

ACT UP New York. n.d. http://www.actupny.org.

Allhoff, F., P. Lin, J. Moor, and J. Weckert, eds. 2007. *Nanoethics: The Ethical and Social Implications of Nanotechnology*. Hoboken, NJ: Wiley-Interscience.

Anderson, E. 1995. "Knowledge, Human Interests, and Objectivity in Feminist Epistemology." *Philosophical Topics* 23: 27–58.

Anderson, E. 2004. "Uses of Value Judgments in Science: A General Argument, with Lessons from a Case Study of Feminist Research on Divorce." *Hypatia* 19: 1–24.

Angell, M. 2004. *The Truth about the Drug Companies: How They Deceive Us and What to Do about It*. New York: Random House.

Basken, P. 2014. "NSF-Backed Scientists Raise Alarm over Congressional Inquiry." *Chronicle of Higher Education*, October 24, A4.

Beder, S. 2000. *Global Spin: The Corporate Assault on Environmentalism*. Rev. ed. White River Junction, VT: Chelsea Green.

Bekelman, J., J. Lee, and C. Gross. 2003. "Scope and Impact of Financial Conflicts of Interest in Biomedical Research." *Journal of the American Medical Association* 289: 454–465.

Berndt, E., and J. Hurvitz. 2005. "Vaccine Advance-Purchase Agreements for Low-Income Countries: Practical Issues." *Health Affairs* 24: 653–665.

Betz, G. 2013. "In Defence of the Value Free Ideal." *European Journal for Philosophy of Science* 3: 207–220.

Biddle, J. 2014a. "Can Patents Prohibit Research? On the Social Epistemology of Patenting and Licensing in Science." *Studies in History and Philosophy of Science* 45: 14–23.

Biddle, J. 2014b. "Intellectual Property in the Biomedical Sciences." In *Routledge Companion to Bioethics*, edited by J. Arras, E. Fenton, and R. Kukla, 149–161. London: Routledge.

Biddle, J., and A. Leuschner. 2015. "Climate Skepticism and the Manufacture of Doubt: Can Dissent in Science Be Epistemically Detrimental?" *European Journal for the Philosophy of Science* 5: 261–278.

Bombardieri, M. 2005. "Summers' Remarks on Women Draw Fire." *Boston Globe*, January 17.

Bombardieri, M. 2006. "Some Seek a Scholar's Return." *Boston Globe*, June 6.

Brigandt, I. 2015. "Social Values Influence the Adequacy Conditions of Scientific Theories: Beyond Inductive Risk." *Canadian Journal of Philosophy* 45: 326–356.

Brown, J. R. 2002. "Funding, Objectivity, and the Socialization of Medical Research." *Science and Engineering Ethics* 8: 295–308.

Brown, M. 2013. "Values in Science beyond Underdetermination and Inductive Risk." *Philosophy of Science* 80: 829–839.

Brown, P., and E. Mikkelsen. 1990. *No Safe Place: Toxic Waste, Leukemia, and Community Action*. Berkeley: University of California Press.

Brulle, R. 2014. "Institutionalizing Delay: Foundation Funding and the Creation of U.S. Climate Change Counter-Movement Organizations." *Climatic Change* 122: 681–694.

Bush, V. 1945. *Science: The Endless Frontier*. Washington, DC: Government Printing Office.

Capek, S. 2000. "Reframing Endometriosis: From 'Career Woman's Disease' to Environment/Body Connections." In *Illness and the Environment: A Reader in Contested Medicine*, edited by S. Kroll-Smith, P. Brown, and V. Gunter, 345–363. New York: New York University Press.

Cho, M. 2006. "Racial and Ethnic Categories in Biomedical Research: There Is No Baby in the Bathwater." *Journal of Law, Medicine & Ethics* 34: 497–499.

Ciarelli, N., and A. Troianovski. 2006. " 'Tawdry Shleifer Affair' Stokes Faculty Anger Toward Summers." *Harvard Crimson*, February 10.

Claudio, L. 1996. "New Asthma Efforts in the Bronx." *Environmental Health Perspectives* 104: 1028–1029.

Claudio, L. 2000. "Reaching Out to New York Neighborhoods." *Environmental Health Perspectives* 108: A450–A451.

Cohn, J. 2006. "The Use of Race and Ethnicity in Medicine: Lessons from the African-American Heart Failure Trial." *Journal of Law, Medicine & Ethics* 34: 552–554.

Colborn, T., D. Dumanoski, and J. P. Myers. 1996. *Our Stolen Future*. New York: Penguin.

Cook, G. 2002. "OncoMouse Breeds Controversy/Cancer Researchers at Odds with DuPont over Fees for Patents." *Boston Globe*, June 2.

Corburn, J. 2005. *Street Science: Community Knowledge and Environmental Health Justice*. Cambridge, MA: MIT Press.

Cranor, C. 1990. "Some Moral Issues in Risk Assessment." *Ethics* 101: 123–143.

Cranor, C. 1993. *Regulating Toxic Substances: A Philosophy of Science and the Law*. New York: Oxford University Press.

Cranor, C. 1995. "The Social Benefits of Expedited Risk Assessments." *Risk Analysis* 15: 353–358.

Cranor, C. 2011. *Legally Poisoned: How the Law Puts Us at Risk from Toxicants*. Cambridge, MA: Harvard University Press.

Crichton, M. 2002. *Prey*. New York: HarperCollins.

Crimp, D. 2011. "Before Occupy: How AIDS Activists Seized Control of the FDA in 1988." *The Atlantic*, December 6.

Dahlberg, K. 1979. *Beyond the Green Revolution: The Ecology and Politics of Global Agricultural Development*. New York: Plenum.

Davies, S., P. Macnaghten, and M. Kearnes. 2009. *Reconfiguring Responsibility: Lessons for Public Policy (Part 1 of the Report on Deepening Debate on Nanotechnology)*. Durham, UK: University of Durham.

De Melo-Martin, I., and K. Intemann. 2016. "The Risk of Using Inductive Risk to Challenge the Value-Free Ideal." *Philosophy of Science* 83: 500–520.

Deparle, J. 1990. "Rude, Rash, Effective, ACT UP Shifts AIDS Policy." *New York Times*, January 3.

Diekmann, S., and M. Peterson. 2013. "The Role of Non-Epistemic Values in Engineering Models." *Science and Engineering Ethics* 19: 207–218.

Douglas, H. 2000. "Inductive Risk and Values in Science." *Philosophy of Science* 67: 559–579.

Douglas, H. 2003. "The Moral Responsibilities of Scientists: Tensions between Autonomy and Responsibility." *American Philosophical Quarterly* 40: 59–68.

Douglas, H. 2007. "Inserting the Public into Science." In *Democratization of Expertise? Exploring Novel Forms of Scientific Advice in Political Decision-Making*, edited by S. Maasen and P. Weingart, 153–169. New York: Springer.

Douglas, H. 2009. *Science, Policy, and the Value-Free Ideal*. Pittsburgh: University of Pittsburgh Press.

Douglas, H. 2015. "Values in Science." *Oxford Handbook of Philosophy of Science*. doi: 10.1093/oxfordhb/9780199368815.013.28.

Dupré, J. 2007. "Fact and Value." In *Value-Free Science? Ideals and Illusions*, edited by H. Kincaid, A. Wylie, and J. Dupre, 27–41. New York: Oxford University Press.

Egan, M. 2007. *Barry Commoner and the Science of Survival*. Cambridge, MA: MIT Press.

Elliott, K. 2009. "The Ethical Significance of Language in the Environmental Sciences: Case Studies from Pollution Research." *Ethics, Place & Environment* 12: 157–173.

Elliott, K. 2011a. "Direct and Indirect Roles for Values in Science." *Philosophy of Science* 78: 303–324.

Elliott, K. 2011b. *Is a Little Pollution Good for You? Incorporating Societal Values in Environmental Research*. New York: Oxford University Press.

Elliott, K. 2013a. "Douglas on Values: From Indirect Roles to Multiple Goals." *Studies in History and Philosophy of Science* 44: 375–383.

Elliott, K. 2013b. "Selective Ignorance and Agricultural Research." *Science, Technology & Human Values* 38: 328–350.

Elliott, K. 2016a. "Climate Geoengineering." In *The Argumentative Turn in Policy Analysis: Reasoning about Uncertainty*, edited by G. Hirsch Hadorn and S. Ove Hansson, 305–324. Dordrecht: Springer.

Elliott, K. 2016b. "Environment." In *Miseducation: A History of Ignorance Making in America and Beyond*, edited by A. J. Angulo, 96–122. Baltimore, MD: Johns Hopkins University Press.

Elliott, K. 2016c. "Science and Policy." In *A Companion to the History of American Science*, edited by M. Largent and G. Montgomery, 468–478. Malden, MA: Wiley-Blackwell.

Elliott, K. 2017. "The Plasticity and Recalcitrance of Wetlands." In *Research Objects in Their Technological Setting*, edited by Bernadette Bensaude-Vincent, Sacha Loeve, Alfred Nordmann, and Astrid Schwarz, forthcoming. London: Pickering & Chatto.

Elliott, K., and D. McKaughan. 2014. "Non-Epistemic Values and the Multiple Goals of Science." *Philosophy of Science* 81: 1–21.

Elliott, K., and D. Resnik. 2014. "Science, Policy, and the Transparency of Values." *Environmental Health Perspectives* 122: 647–650.

Elliott, K., and D. Steel, eds. 2017. *Current Controversies in Science and Values*. New York: Routledge.

Elliott, K., and D. Willmes. 2013. "Cognitive Attitudes and Values in Science." *Philosophy of Science* 80 (2013 Proceedings): 807–817.

Epstein, S. 1996. *Impure Science: AIDS, Activism, and the Politics of Knowledge*. Berkeley: University of California Press.

Fehr, C. 2011. "Feminist Philosophy of Biology." *Stanford Encyclopedia of Philosophy*. http://plato.stanford.edu/entries/feminist-philosophy-biology.

Frickel, S., S. Gibbon, J. Howard, J. Kempner, G. Ottinger, and D. Hess. 2010. "Undone Science: Charting Social Movement and Civil Society Challenges to Research Agenda Setting." *Science, Technology & Human Values* 35: 444–473.

Frodeman, R., J. B. Holbrook, P. Bourexis, S. Cook, L. Diderick, and R. Tankersley. 2013. "Broader Impacts 2.0: Seeing—and Seizing—the Opportunity." *BioScience* 63: 153–154.

Funtowicz, S., and J. Ravetz. 1992. "Three Types of Risk Assessment and the Emergence of Post-Normal Science." In *Social Theories of Risk*, edited by S. Krimsky and D. Golding, 251–274. Westport, CT: Praeger.

Gardiner, S. 2004. "Ethics and Global Climate Change." *Ethics* 114: 555–600.

Gardner, H. 1995. "Scholarly Brinkmanship." In *The Bell Curve Debate: History, Documents, Opinions*, edited by R. Jacoby and N. Glauberman, 61–72. New York: Times Books.

Gillis, J., and J. Schwartz. 2015. "Deeper Ties to Corporate Cash for Doubtful Climate Researcher." *New York Times*, February 21.

Gobster, P. 1997. "The Chicago Wilderness and Its Critics III: The Other Side A Survey of the Arguments." *Restoration & Management Notes* 15: 32–37.

Goldacre, B. 2012. *Bad Pharma: How Drug Companies Mislead Doctors and Harm Patients*. New York: Faber and Faber.

Gordin, M. 2012. "How Lysenkoism Became Pseudoscience: Dobzhansky to Velikovsky." *Journal of the History of Biology* 45: 443–468.

Gotzsche, P. 2013. *Deadly Medicines and Organised Crime: How Big Pharma Has Corrupted Healthcare*. London: Radcliffe.

Gould, S. 1981. *The Mismeasure of Man*. New York: Norton.

Gould, S. 1995. "Mismeasure by Any Measure." In *The Bell Curve Debate: History, Documents, Opinions*, edited by R. Jacoby and N. Glauberman, 3–13. New York: Times Books.

Graham, L. 1993. *Science in Russia and the Soviet Union*. Cambridge: Cambridge University Press.

Greenberg, D. 1968. *The Politics of Pure Science*. New York: New American Library.

Grens, K. 2015. "Chronic Fatigue Syndrome Reframed." *The Scientist*, February 11.

Guston, D. 2000. *Between Politics and Science: Assuring the Integrity and Productivity of Research*. Cambridge: Cambridge University Press.

Guston, D. 2008. "Innovation Policy: Not Just a Jumbo Shrimp." *Nature* 454: 940–941.

Guston, D. 2014. "Building the Capacity for Public Engagement with Science in the United States." *Public Understanding of Science* 23: 53–59.

Harding, S. 2015. *Objectivity and Diversity: Another Logic of Scientific Research*. Chicago: University of Chicago Press.

Harr, J. 1995. *A Civil Action*. New York: Random House.

Healy, D., and D. Catell. 2003. "Interface between Authorship, Industry, and Science in the Domain of Therapeutics." *British Journal of Psychiatry* 183: 22–27.

Heaney, C., S. Wilson, and O. Wilson. 2007. "The West End Revitalization Association's Community-Owned and -Managed Research Model: Development, Implementation, and Action." *Progress in Community Health Partnerships: Research, Education, and Action* 1: 339–349.

Herrnstein, R., and C. Murphy. 1994. *The Bell Curve: Intelligence and Class Structure in American Life*. New York: Free Press.

Hicks, D. 2014. "A New Direction for Science and Values." *Synthese* 191: 3271–3295.

Holman, B. 2015. "The Fundamental Antagonism: Science and Commerce in Medical Epistemology." PhD diss., University of California–Irvine.

Hough, P., and M. Robertson. 2009. "Mitigation under Section 404 of the Clean Water Act: Where It Comes from, What It Means." *Wetlands Ecology and Management* 17: 15–33.

Hudson, R. 2016. "Why We Should Not Reject the Value-Free Ideal of Science." *Perspectives on Science* 24: 167–191.

IAASTD (International Assessment of Agricultural Knowledge, Science, and Technology for Development). 2009. *Agriculture at a Crossroads: Executive Summary of the Synthesis Report.* Washington, DC: Island Press.

Intemann, K. 2015. "Distinguishing between Legitimate and Illegitimate Values in Climate Modeling." *European Journal for Philosophy of Science.* doi: 10.1007/s13194-014-0105-6.

Irwin, A. 1995. *Citizen Science: A Study of People, Expertise, and Sustainable Development.* London: Routledge.

Irwin, A. 2001. "Constructing the Scientific Citizen: Science and Democracy in the Biosciences." *Public Understanding of Science* 10: 1–18.

Jasanoff, S., and B. Wynne. 1998. "Science and Decision Making." In *Human Choice and Climate Change*, vol. 1, edited by S. Rayner and E. Malone, 1–87. Columbus, OH: Battelle Press.

John, S. 2015. "Inductive Risk and the Contexts of Communication." *Synthese* 192: 79–96.

Kahan, D. 2010. "Fixing the Communications Failure." *Nature* 463: 296–297.

Kaime, E., K. Moore, and S. Goldberg. 2010. "CDMRP: Fostering Innovation through Peer Review." *Technology and Innovation* 12: 233–240.

Kamin, L. 1995. "Lies, Damned Lies, and Statistics." In *The Bell Curve Debate: History, Documents, Opinions*, edited by R. Jacoby and N. Glauberman, 81–105. New York: Times Books.

Kassirer, J. 2005. *On the Take: How Medicine's Complicity with Big Business Endangers Your Health.* New York: Oxford University Press.

Katikireddi, S. V., and S. Valles. 2015. "Coupled Ethical-Epistemic Analysis of Public Health Research and Practice: Categorizing Variables to Improve Population Health and Equity." *American Journal of Public Health* 105: e36–e42.

Kavanagh, E., ed. 2007. "The Risks and Advantages of Framing Science." *Science* 317: 1168–1169.

Kerr, R. 1989. "Hansen vs. the World on the Greenhouse Threat." *Science* 244: 1041–1043.

Kessler, G. 2015. "Setting the Record Straight: The Real Story of a Pivotal Climate-Change Hearing." *Washington Post*, March 30.

Kinchy, A. 2010. "Epistemic Boomerang: Expert Policy Advice as Leverage in the Campaign against Transgenic Maize in Mexico." *Mobilization: An International Journal* 15: 179–198.

Kintisch, E. 2014. "Should the Government Fund Only Science in the 'National Interest'?" *National Geographic News*, October 29.

Kitcher, Philip. 1990. "The Division of Cognitive Labor." *Journal of Philosophy* 87: 5–22.

Kitcher, P. 2001. *Science, Truth, and Democracy.* New York: Oxford University Press.

Koertge, N. 1998. *A House Built on Sand: Exposing Postmodernist Myths about Science.* New York: Oxford University Press.

Kourany, J. 2010. *Philosophy of Science after Feminism.* New York: Oxford University Press.

Krimsky, S. 2000. *Hormonal Chaos: The Scientific and Social Origins of the Environmental Endocrine Hypothesis.* Baltimore, MD: Johns Hopkins University Press.

Krimsky, S. 2003. *Science in the Private Interest: Has the Lure of Profits Corrupted Biomedical Research?* Lanham, MD: Rowman and Littlefield.

Kupferschmidt, K. 2015. "Rules of the Name." *Science* 348: 745.

Lacey, H. 1999. *Is Science Value Free?* London: Routledge.

Largent, M. 2012. *Vaccine: The Debate in Modern America.* Baltimore, MD: Johns Hopkins University Press.

Larson, B. 2011. *Metaphors for Environmental Sustainability: Redefining our Relationships with Nature.* New Haven, CT: Yale University Press.

Lave, R. 2009. "The Controversy over Natural Channel Design: Substantive Explanations and Potential Avenues for Resolution." *Journal of the American Water Resources Association* 45: 1519–1532.

Leonhardt, D. 2007. "A Battle over the Costs of Global Warming." *New York Times*, February 21.

Lim, M., Z. Wang, D. Olazabal, X. Ren, E. Terwilliger, and L. Young. 2004. "Enhanced Partner Preference in a Promiscuous Species by Manipulating the Expression of a Single Gene." *Nature* 429: 754–757.

Lok, C. 2010. "Science Funding: Science for the Masses." *Nature* 465: 416–418.

Longino, H. 1990. *Science as Social Knowledge.* Princeton, NJ: Princeton University Press.

Longino, H. 2002. *The Fate of Knowledge.* Princeton, NJ: Princeton University Press.

Lovejoy, C. O. 1981. "The Origin of Man." *Science* 211: 341–350.

Ludwig, D. 2016. "Ontological Choices and the Value-Free Ideal." *Erkenntnis* forthcoming. doi: 10.1007/s10670-015-9793-3.

Lyon, S., J. A. Bezaury, and T. Mutersbaugh. 2010. "Gender Equity in Fair Trade-Organic Coffee Producer Organizations: Cases from Mesoamerica." *Geoforum* 41: 93–103.

MacLean, D. 2009. "Environmental Ethics and Future Generations." In *Nature in Common? Environmental Ethics and the Contested Foundations of Environmental Policy*, edited by Ben Minteer, 118–141. Philadelphia: Temple University Press.

Malakoff, D. 2004. "The River Doctor." *Science* 305: 937–939.

McCright, A., and R. Dunlap. 2010. "Anti-Reflexivity: The American Conservative Movement's Success in Undermining Climate Science and Policy." *Theory, Culture & Society* 27: 100–133.

McKaughan, D. 2012. "Voles, Vasopressin, and Infidelity: A Molecular Basis for Monogamy, a Platform for Ethics, and More?" *Biology and Philosophy* 27: 521–543.

McKaughan, D., and K. Elliott. 2013. "Backtracking and the Ethics of Framing: Lessons from Voles and Vasopressin." *Accountability in Research* 20: 206–226.

Merton, R. 1942. *The Sociology of Science.* Chicago: University of Chicago Press.

Mervis, J. 2014. "Congress, NSF Spar on Access to Grant Files." *Science* 346: 152–153.

Michaels, D. 2008. *Doubt Is Their Product: How Industry's Assault on Science Threatens Your Health.* New York: Oxford University Press.

Musschenga, A., W. van der Steen, and V. Ho 2010. "The Business of Drug Research: A Mixed Blessing." In *The Commodification of Academic Research*, edited by H. Radder, 110–131. Pittsburgh: University of Pittsburgh Press.

Nabhan, G. P. 2009. *Where Our Food Comes From: Retracing Nikolay Vavilov's Quest to End Famine.* Washington, DC: Island Press.

NAS (National Academies of Sciences, Engineering, and Medicine). 2016. *Genetic Engineered Crops: Experiences and Prospects.* Washington, DC: National Academies Press.

Niemitz, C. 2010. "The Evolution of the Upright Posture and Gait—A Review and a New Synthesis." *Naturwissenschaften* 97: 241–263.

Nisbet, M., and C. Mooney. 2007. "Science and Society. Framing Science." *Science* 316: 56.

Noble, H. 1999. "Far More Poor Children are Hospitalized for Asthma, Study Shows." *New York Times*, July 27.

Nordhaus, W. 2007. "Critical Assumptions in the Stern Review on Climate Change." *Science* 317: 201–202.

NRC (National Research Council). 1995. *Wetlands: Characteristics and Boundaries.* Washington, DC: National Academy Press.

NRC (National Research Council). 1996. *Understanding Risk: Informing Decisions in a Democratic Society.* Washington, DC: National Academies Press.

NRC (National Research Council). 1999. *Hormonally Active Agents in the Environment.* Washington, DC: National Academies Press.

NRC (National Research Council). 2015. *Climate Intervention: Reflecting Sunlight to Cool Earth.* Washington, DC: National Academies Press.

O'Fallon, L., and S. Finn. 2015. "Citizen Science and Community-Engaged Research in Environmental Health." *Lab Matters* Fall: 5.

Okruhlik, K. 1994. "Gender and the Biological Sciences." *Canadian Journal of Philosophy* Supplementary vol. 20: 21–42.

Oreskes, N., and E. Conway. 2010. *Merchants of Doubt: How a Handful of Scientists Obscured the Truth on Issues from Tobacco Smoke to Global Warming.* New York: Bloomsbury.

O'Rourke, M., and S. Crowley. 2013. "Philosophical Intervention and Cross-Disciplinary Science: The Story of the Toolbox Project." *Synthese* 190: 1937–1954.

Ottinger, G. 2010. "Buckets of Resistance: Standards and the Effectiveness of Citizen Science." *Science, Technology & Human Values* 35: 244–270.

Park, P. 2004. "EPO Restricts OncoMouse Patent." *The Scientist*, July 26.

Parker, W. 2009. "Confirmation and Adequacy-for-Purpose in Climate Modelling." *Aristotelian Society Supplementary Volume* 83: 233–249.

Parthasarathy, S. 2007. *Genetic Medicine: Breast Cancer, Technology, and the Comparative Politics of Health Care.* Cambridge, MA: MIT Press.

Parthasarathy, S. 2011. "Whose Knowledge? What Values? The Comparative Politics of Patenting Life Forms in the United States and Europe." *Policy Sciences* 44: 267–288.

Patel, R. 2007. *Stuffed and Starved: The Hidden Battle for the World's Food System.* Brooklyn, NY: Melville House.

Perfecto, I., J. Vandermeer, and A. Wright. 2009. *Nature's Matrix: Linking Agriculture, Conservation, and Food Sovereignty.* London: Earthscan.

Pielke, R., Jr. 2007. *The Honest Broker: Making Sense of Science in Policy and Politics.* Cambridge: University of Cambridge Press.

Pielke, R., Jr. 2010. *The Climate Fix: What Scientists and Politicians Won't Tell You about Global Warming.* New York: Basic Books.

Pierrehumbert, R. 2015. "Climate Hacking Is Barking Mad." *Slate*, February 10.

Plotz, David. 2001. "Larry Summers: How the Great Brain Learned to Grin and Bear It." *Slate*, June 29.

Pogge, T. 2009. "The Health Impact Fund and Its Justification by Appeal to Human Rights." *Journal of Social Philosophy* 40: 542–569.

"Police Arrest AIDS Protesters Blocking Access to FDA Offices." 1988. *Los Angeles Times*, October 11.

Potochnik, A. 2015. "The Diverse Aims of Science." *Studies in History and Philosophy of Science* 53: 71–80.

Potrykus, I. 2002. "Golden Rice and the Greenpeace Dilemma." In *Genetically Modified Foods*, edited by M. Ruse and D. Castle, 55–57. Amherst, NY: Prometheus.

Powell, M., and J. Powell. 2011. "Invisible People, Invisible Risks: How Scientific Assessments of Environmental Health Risks Overlook Minorities—and How Community Participation Can Make Them Visible." In *Technoscience and Environmental Justice: Expert Cultures in a Grassroots Movement*, edited by G. Ottinger and B. Cohen, 149–178. Cambridge, MA: MIT Press.

Pringle, P. 2003. *Food Inc.: Mendel to Monsanto—The Promises and Perils of the Biotech Industry*. New York: Simon and Schuster.

Pringle, P. 2008. *The Murder of Nikolai Vavilov*. New York: Simon and Schuster.

Proctor, R. 1991. *Value-Free Science? Purity and Power in Modern Knowledge*. Cambridge, MA: Harvard University Press.

Proctor, R. 2012. *Golden Holocaust: Origins of the Cigarette Catastrophe and the Case for Abolition*. Berkeley: University of California Press.

Raffensperger, C., and J. Tickner, eds. 1999. *Protecting Public Health and the Environment*. Washington, DC: Island Press.

Reiss, J., and P. Kitcher. 2009. "Biomedical Research, Neglected Diseases, and Well-Ordered Science." *Theoria* 66: 263–282.

Robertson, M. 2006. "The Nature that Capital Can See: Science, State, and Market in the Commodification of Ecosystem Services." *Environment and Planning D: Society and Space* 24: 367–387.

Roll-Hansen, N. 1985. "A New Perspective on Lysenko?" *Annals of Science* 42: 261–278.

Rose, S. 2009. "Darwin 200: Should Scientists Study Race and IQ? No: Science and Society Do Not Benefit." *Nature* 457: 786–788.

Sarewitz, D. 1996. *Frontiers of Illusion: Science, Technology, and the Politics of Progress*. Philadelphia: Temple University Press.

Sarewitz, D. 2007. "How Science Makes Environmental Controversies Worse." *Environmental Science & Policy* 7: 385–403.

Schienke, E. W., S. D. Baum, N. Tuana, K. J. Davis, and K. Keller. 2011. "Intrinsic Ethics Regarding Integrated Assessment Models for Climate Management." *Science and Engineering Ethics* 17: 503–523.

Schlesinger, S., and S. Kinzer. 1999. *Bitter Fruit: The Story of the American Coup in Guatemala*. Cambridge, MA: Harvard University Press.

Schuurbiers, D., and E. Fisher. 2009. "Lab-Scale Intervention." *EMBO Reports* 10: 424–427.

Shabecoff, P. 1988. "Global Warming Has Begun, Expert Tells Senate." *New York Times*, June.

Shah, S. 2010. *The Fever: How Malaria Has Ruled Humankind for 500,000 Years*. New York: Farrar, Straus, and Giroux.

Shiva, V. 1988. "Reductionist Science as Epistemological Violence." In *Science, Hegemony, and Violence: A Requiem for Modernity*, edited by A. Nandy, 232–256. Oxford: Oxford University Press.

Shiva, V. 2002. "Golden Rice Hoax: When Public Relations Replace Science." In *Genetically Modified Foods*, edited by M. Ruse and D. Castle, 58–62. Amherst, NY: Prometheus.

Shore, D. 1997. "The Chicago Wilderness and Its Critics II: Controversy Erupts over Restoration in Chicago Area." *Restoration & Management Notes* 15: 25–31.

Shrader-Frechette, K. 1991. *Risk and Rationality: Philosophical Foundations for Populist Reforms*. Berkeley: University of California Press.

Shrader-Frechette, K. 1996. *The Ethics of Scientific Research.* Lanham, MD: Rowman and Littlefield.

Shrader-Frechette, K. 2007. *Taking Action, Saving Lives: Our Duties to Protect Environmental and Public Health.* New York: Oxford University Press.

Shrader-Frechette, K. 2014. *Tainted: How Philosophy of Science Can Expose Bad Science.* New York: Oxford University Press.

Sismondo, S. 2007. "Ghost Management: How Much of the Medical Literature is Shaped Behind the Scenes by the Pharmaceutical Industry?" *PLoS Medicine* 4:e286.

Sismondo, S. 2008. "Pharmaceutical Company Funding and Its Consequences: A Qualitative Systematic Review." *Contemporary Clinical Trials* 29: 109–113.

Smith, H. 2014. "Remembering the Genius Who Got BPA Out of Your Water Bottles, and So Much More." *Grist*, December 16.

Solomon, M. 2001. *Social Empiricism.* Cambridge, MA: MIT Press.

Sonderholm, J. 2010. "A Theoretical Flaw in the Advance Market Commitment Idea." *Journal of Medical Ethics* 36: 339–343.

Steel, D. 2010. "Epistemic Values and the Argument from Inductive Risk." *Philosophy of Science* 77: 14–34.

Stern, N. 2007. *The Economics of Climate Change: The Stern Review.* Cambridge: Cambridge University Press.

Stokstad, E. 2008. "Dueling Visions for a Hungry World." *Science* 319 (14 March): 1474–1476.

Tickner, J. 1999. "A Map Toward Precautionary Decision Making." In *Protecting Public Health & the Environment: Implementing the Precautionary Principle*, edited by C. Raffesnperger and J. Ticker, 162–186. Washington, DC: Island Press.

Toumey, C. 2006. Science and Democracy. *Nature Nanotechnology* 1: 6–7.

Tuana, N. 2010. "Leading with Ethics, Aiming for Policy: New Opportunities for Philosophy of Science." *Synthese* 177: 471–492.

Tuana, N., R. Sriver, T. Svogoda, R. Olson, P. Irvine, J. Haqq-Misra, and K. Keller. 2012. "Towards Integrated Ethical and Scientific Analysis of Geoengineering: A Research Agenda." *Ethics, Policy & Environment* 15: 136–157.

Tucker, A. 2014, "What Can Rodents Tell Us about Why Humans Love." *Smithsonian Magazine*, February.

Tuller, D. 2007. "Chronic Fatigue No Longer Seen as 'Yuppie Flu.'" *New York Times*, July 17.

Turner, E., A. Matthews, E. Linardatos, R. Tell, and R. Rosenthal. 2008. "Selective Publication of Antidepressant Trials and Its Influence on Apparent Efficacy." *New England Journal of Medicine* 358: 252–260.

Uscinski, J., and C. Klofstad. 2010. "Who Likes Political Science?: Determinants of Senators' Votes on the Coburn Amendment." *Political Science & Politics* 43: 701–706.

Wade, N. 2011. "Scientists Measure the Accuracy of a Racism Claim." *New York Times*, June 13.

Walum, H., L. Westberg, S. Henningsson, J. Neiderhiser, D. Reiss, W. Igl, J. Ganiban, et al. 2008. "Genetic Variation in the Vasopressin Receptor 1a Gene (AVPR1A) Associates with Pair-Bonding Behavior in Humans." *Proceedings of the National Academy of Sciences* 105: 14153–14156.

Weart, S. 2014. "The Public and Climate Change." http://www.aip.org/history/climate/public2.htm.

Weitzman, M. 2007. "A Review of the *Stern Review* on the Economics of Climate Change." *Journal of Economic Literature* 45: 703–724.

Wickson, F., and B. Wynne. 2012. "Ethics of Science for Policy in the Environmental Governance of Biotechnology: MON810 Maize in Europe." *Ethics, Policy & Environment* 15: 321–340.

Wilholt, T. 2009. "Bias and Values in Scientific Research." *Studies in History and Philosophy of Science* 40: 92–101.

Wilholt, T. 2013. "Epistemic Trust in Science." *British Journal for Philosophy of Science* 64: 233–253.

Wylie, A. 1996. "The Constitution of Archaeological Evidence: Gender Politics and Science." In *The Disunity of Science: Boundaries, Contexts, and Power*, edited by P. Galison and D. Stump, 311–343. Stanford, CA: Stanford University Press.

Wynne, B. 1992. "Public Understanding of Science Research: New Horizons of Hall of Mirrors?" *Public Understanding of Science* 1: 37–43.

Wynne, B. 2005. "Risk as Globalising 'Democratic' Discourse? Framing Subjects and Citizens." In *Science and Citizens: Globalization and the Challenge of Engagement*, edited by M. Leach, I. Scoones, and B. Wynne, 66–82. London: Zed Books.

Ye, X., S. Al-Babili, A. Kloti, J. Zhang, P. Lucca, P. Beyer, and I. Potrykus. 2000. "Engineering the Provitamin A (β-Carotene) Biosynthetic Pathway into (Carotenoid-Free) Rice Endosperm." *Science* 287: 303–305.

Young, L. 2009. "Being Human: Love—Neuroscience Reveals All." *Nature* 457: 148.

Zihlman, A. 1985. "Gathering Stories for Hunting Human Nature." *Feminist Studies* 11: 365–377.

INDEX

community-based participatory research
(CBPR), 16–17, 49–50, 60, 81,
99, 137–144, 158, 160, 175–176.
See also community owned and
managed research; engagement;
popular epidemiology; public
participation
confirmation bias, 22, 102, 130
Congress, 25–31, 39, 83–87, 101,
171, 180
consensus conferences, 147–148
Conservation Foundation, 5
constitutive values. *See* values: epistemic
contextual values. *See*
values: nonepistemic
Conway, Erik, 100–102, 107
Conway, Gordon, 44
co-production, 167, 177
Córdova, France, 26
corporations. *See* industry
Cosmopolitan, 139
Cranor, Carl, 57, 64–66, 68, 80, 88,
99, 156
creationism. *See* intelligent design
Crichton, Michael, 145
Crowley, Stephen, 152–153
curiosity, 25
Current Medical Directions, 54
Cuvier, Georges, 129

Damasio, Antonio, 115
Darwin, Charles, 129
Darwinism. *See* evolution
data, x, 14, 87, 117, 128–129, 150, 171.
See also evidence
DDT, 33
deficit model, 169–170, 177, 184. *See
also* local knowledge; science policy
deliberation. *See* engagement
democracy, 164, 176. *See also*
communism; values: political
democratizing science. *See* engagement
Denmark, 147
Department of Agriculture, 26, 29
Department of Defense, 26, 29, 93,
102, 180
Department of Energy (DOE), 26, 29,
101, 150–151
depression, 53–56
Diamond v. Chakrabarty, 155

Diekmann, Sven, 78–79
dioxin, 92–99, 165
disclosure. *See* transparency
discounting, 75–79, 81. *See also*
economics; models
discrimination, 23. *See also* gender; race
disinterestedness, 8
dissemination (of scientific information).
See science communication
Divine Comedy, 66
DNA, 158. *See also* genetics
dose response, 6, 125–127, 133–135
hormetic, 126, 133–135
nonmonotonic, 6, 127, 133–135
threshold, 125
Doubt Is Their Product, ix, 102
Douglas, Heather, x, xiii, 92–99, 108,
167, 182
Dow Chemical Company, 96. *See also*
industry: chemical
drug industry. *See*
industry: pharmaceutical
DuPont, 152
Dupré, John, x, 124

ecology, 45–48, 67–68, 86–89, 120–122.
See also agroecology
economics, 20, 45–48, 59, 65–66, 75–79,
116–117. *See also* values: economic
Egan, Michael, 165
Eli Lilly, 55
Elliott, Janelle, xiii
Elliott, Janet, xiv
Elliott, Jayden, xiv
Elliott, Kevin, 118–119
Elliott, Leah, xiv
embryonic stem-cell research. *See* stem-
cell research
The Endless Frontier, 27
endocrine disruption, 5–7, 9–10, 12, 16,
88, 92, 125, 127, 133–135
definition of, 125
and hormonally active agents, 10,
125, 133–135
in humans vs. wildlife, 86
terminology for, 10, 125, 127,
133–135
value judgments concerning, 86–88,
125, 127, 133–135
See also Colborn, Theo

traditional ecological knowledge (TEK). *See* knowledge: traditional ecological

transparency, x, 3, 7, 10–11, 14–15, 17, 53, 74, 93, 99–100, 105–106, 108, 118–120, 167, 170–174, 177–179, 185. *See also* backtracking; engagement; representativeness

Travolta, John, 49

Truman, Harry, 27

trust, 85, 107, 114–115

truth, 62, 64, 71–73, 76, 81, 120

TSCA. *See* Toxic Substances Control Act

Tuana, Nancy, x, 78, 81, 151–152, 184

Type I error. *See* false positives

Type II error. *See* false negatives

uncertainty, 6, 9, 75–76, 83–105, 108, 116, 134, 148, 151–152, 183
  communication of, 83–99, 134
  manufacture of, 100–103, 107–108, 171, 183
  responding to, 103–105
  See also *Doubt Is Their Product*; *Merchants of Doubt*

underdetermination, x

understanding, of information, 5, 30, 51, 55, 85, 88–92, 99, 112–120, 123–124, 127–128, 133, 164, 169. *See also* science communication; value judgments: in communicating results

unicorn, xi

United Kingdom, 75, 146

United Nations Development Program, 44

United Nations Environment Program, 44

United States, 44, 47, 57, 59, 62, 66–67, 79, 85, 157, 160

universities, 45, 153

University of California, Berkeley, 61, 63

University of California, Los Angeles (UCLA), 114

University of Maryland, 85

University of Massachusetts, Amherst, 126

University of Miami, 29

University of Notre Dame, 149

University of Toronto, 85

vaccines, 8, 91, 103–104, 107, 109, 119, 169, 182

Valentine's Day, 111

Valles, Sean, 131

value judgments, x, 7–60, 83–99, 111–135, 142, 145–146, 152–153, 167, 170–172, 177
  in choosing research questions and methods, 7–8, 12, 16, 41–60, 142, 150, 152–153, 160, 166, 174, 177
  in choosing study designs, x, 15, 142, 176–177
  in choosing topics, x, 7–8, 19–38, 142, 148, 166, 174, 177
  in communicating results, 15, 83–92, 111–135, 142, 148, 152, 166, 174, 177, 183
  definition of, 12
  in developing language and terminology, x, 7–8, 10, 12–13, 16, 120–128, 134–135, 142, 166, 183 (*see also* language; metaphor)
  in evaluating and interpreting studies, 12, 22, 49–50, 142
  individual vs institutional, 12, 37
  intentional vs unintentional, 12–13, 72, 108, 155–156, 177
  in making assumptions, 12, 15–16, 41, 48–53, 58–59, 96–97, 149–151, 153, 160, 166, 171
  problems caused by, 7
  in responding to uncertainty, x, 9, 83–105, 166, 177
  in setting standards of evidence, 7–8, 16, 83, 92–99, 142, 150, 174, 182
  in setting the aims of research, x, 8–9, 61–81, 142, 152, 166, 177

value neutrality, 13, 111–112, 117, 122–123, 133, 150